Soulful

# Soulful

## YOU IN THE FUTURE
## OF ARTIFICIAL INTELLIGENCE

David Espindola

 Brainyus

Soulful: You in the Future of Artificial Intelligence
© 2023 David Espindola

Limit of Liability / Disclaimer of Warranty

ISBN: 979-8-9878742-0-2 (paperback)
ISBN: 979-8-9878742-1-9 (hardcover)
ISBN: 979-8-9878742-2-6 (e-book)
Library of Congress Control Number: 2023904381

Cover Design: Brainyus and Paul Nylander | Illustrada
Cover Image: DALL-E 2
Book Design: Paul Nylander | Illustrada

Published by Brainyus, Minneapolis, MN
brainyus.com

*For the Lord, who guides me*
*and sustains me, and my family*
*whom I love dearly.*

# Contents

# Preface

Why would anyone combine the words "Soulful" and "Artificial Intelligence" in a book title? They seem like disparate subjects that hardly fit in a sentence together. In 1982, in his classic book *Megatrends*,[1] author John Naisbitt had the foresight to portray a future that included what he called "high-tech/high-touch" as a megatrend. Naisbitt was on to something significant that is applicable today and will continue to impact our future as humans bring our high-touch capabilities to collaborate with high-tech intelligent machines.

Writing a book covering such contrasting subjects—the soft, intuitive, emotional, high-touch human condition and the hard, analytical, nerdish, high-tech Artificial Intelligence (AI) domain—while keeping it coherent and easy to read is an arduous task. However, it is paramount that we become comfortable dealing with these complexities and juxtapositions as they will continue to be prevalent in our world.

Rest easy if you are nontechnical and hesitant to read a book about Artificial Intelligence. You don't need any technical knowledge to read and understand this book. However, not everyone will be familiar with some of the terms used. Therefore, at the end of the book I provide a Glossary to assist in your comprehension. By the same token, if you are a technical person but hesitate to read a book titled *Soulful*, I think you will find the discussion intriguing and well researched.

Artificial Intelligence has been in development for many decades, but until recently, it was primarily kept within the confines of Computer Science departments and specialized companies. However, in the last few years, it has gained the attention of technologists and managers, and now it has become easily accessible to the masses. In terms of exponential technologies, we have hit an inflection point. Combining the acumen of human intuition with the computational strength of the machines opens the possibility of a new level of productivity unlike any we have ever seen.

I have several goals for this book: to help readers understand the magnitude of the changes Artificial Intelligence will bring; to prepare readers to gain the skills and insights necessary to do well as AI starts to encroach on every aspect of our professions, economy, and lives; to paint an optimistic view of a possible abundant future balanced with a realistic view of the challenges we will face getting there; and finally, to help us envision a way to live with purpose, peace, and prosperity in a world shared with intelligent machines.

To achieve these goals, I have organized the book into the following sections: In chapter 1, I present the subjects covered throughout the book and provide an overview. In chapters 2 and 3, I describe AI's impact in many areas of our lives and how we can collaborate with AI to bring many benefits to humanity. In chapters 4 and 5, I explain how humans are unique and distinct from intelligent machines and how we can acquire new skills to be effective in a human-machine collaborative effort to create a better world.

In chapters 6 and 7, I present an optimistic vision of a possible post-scarcity future contrasted with an earnest view of our reality. A world of abundance and low cost of living is within reach with the technologies available today given their foreseeable advancement rate. However, we are unprepared for this journey, and the current socioeconomic structure is inadequate for our future needs. We must cross the chasm separating our reality from a potentially bright future. If we fail, we could experience collapse. But I offer suggestions from change management principles on how to get across to the other side successfully. Being realistic about our reality does not mean we need to become pessimists. The news media creates a doomsday perception to get our attention. This elevates bad news disproportionately to the good news surrounding us, which remains mostly invisible. We need to be aware of this trend and cautious against its threat.

Chapter 8 discusses what we might find when we get to the other side. I paint a picture of a potential scenario in which machines do all the soulless work to provide for our essential needs. Humans can then turn to more enjoyable and purposeful pursuits that fulfill the needs of our souls.

In writing this book, I used a skill that I explain and propose as one of the most valuable skills as we prepare to collaborate with intelligent machines—lateral thinking. I researched several domains, including technology, neuroscience, social psychology, economics, philosophy, theology, and others to find common threads that pertain to the advancement of Artificial Intelligence and its impact on humanity. I am not an expert in these domains except perhaps technology, where I spent decades in my career. But even in technology, which is a vast field, my experience was mostly managing it rather than developing it. Yet, I can connect the dots across these domains to create a compelling and cohesive argument and convey a thoughtful narrative with a logical conclusion. I believe this is a critical ability that will allow humans to collaborate successfully with machines that have much greater specialized knowledge than any of us could ever hope to attain.

I have referenced and quoted many authors and experts across these domains that I encountered during my research for the book. If I can capture and share a glimpse of the future, it is because I stand humbly on the shoulders of these giants. If you are interested in further researching the many ideas discussed here, I encourage you to take advantage of the references provided.

I would be disingenuous if I did not use AI in a book that portrays the benefits of human-AI collaboration. Thus, I collaborated with DALL-E 2 to create the book cover's image. However, I did not use any text from any Generative AI tools like ChatGPT. All the words used in this book are my own or have been appropriately quoted or credited to the original writers.

I want to acknowledge that not everyone reading this book will agree with my point of view. The book discusses future scenarios that are hypothetical and impossible to predict accurately. However, considering such scenarios may open your mind to possibilities you may not have considered before. In one chapter, I tried to explain how cognitive biases may influence how we form our views and opinions. This applies to you but also to me. On that note, I want to be completely transparent about the fact that I am a follower of Jesus Christ, and as such, my beliefs and viewpoints are based on the foundations of the Christian faith. I also believe that science and faith are not dichotomous but complementary. You may have a different view, and if that is the case, I ask that you consider that reading opposing views with an open mind, despite being difficult, can help us broaden our horizons and be more compassionate. Empathy and compassion, as we will discuss throughout this book, will become increasingly valuable skills when collaborating with AI.

Finally, I would like to thank you for taking the time to read the book. I would appreciate hearing from you and getting your thoughts on it. I can be reached on my website DavidEspindola.com.

# CHAPTER 1:

# Machines Don't Have Souls

"I am not writing another book." A best-selling author, who had already written nine books and was on his way to his tenth, expressed his reluctance on social media. His reasoning? "ChatGPT has arrived and is commoditizing the use of language and words."[2] ChatGPT is an Artificial Intelligence (AI) application that falls under the Generative AI category. GPT stands for Generative Pre-trained Transformer, a geeky name for a technology that is taking the world by storm. Generative AI enables computers to create new compositions using previously created content, such as text, audio, video, images, and code. ChatGPT is not the first AI app of its kind—others like copy.ai and jasper.ai have been around a bit longer—but ChatGPT takes human-machine collaboration to a whole new level, as it interacts with humans in an almost science fiction sort of way. Researchers have shown that it can even perform Theory of Mind tasks—the innate human ability to infer the thoughts of other humans—at the level of a nine-year-old.[3]

In late 2022, ChatGPT gained popularity with unprecedented speed, reaching over one million users in just five days. It seemed as if we had reached an inflection point in AI advancement and its ease of use, removing any trepidations that novice users might have about interacting with it. The interface is simple, like a Google search, yet much more powerful. You can type any question or request, and ChatGPT returns an astounding response in text format, from a few paragraphs to fully developed essays. Therein lies the fear that the best-selling author expressed on social media that the written word was about to be wholly commoditized, and the value of books would quickly approach zero. But the fear doesn't stop there. What else will AI commoditize? Will human work be entirely replaced by machines? The Luddites first raised this concern in the early days of the Industrial Revolution, and we could argue there is nothing new here. But we could counterargue that we are now dealing with a different beast—machines that could potentially be just as intelligent, if not more, than we are.

A group of expert panelists at the 2023 World Economic Forum in Davos, Switzerland, unanimously agreed that 2023 marks an inflection point for AI. Brad Smith, Vice-Chair and President at Microsoft, explained why:

> Think about the history of technology, and there are certain inflection points when technology is really embraced by the public, and frankly, life is not the same again. The most recent was probably 2007 when the introduction of the iPhone transformed the movement towards mobility.

He then made the case that the same principle applies to AI in 2023:

> The AI has existed, it has been a topic of discussion now for six or seven years at Davos, but this is the year. I do believe 2023 will be the year that is remembered as the inflection point because these large language or foundational models are enabling people to do things they didn't believe would be possible even really in this decade, and it is going to be used in so many ways.

The panel discussed several opportunities and risks that AI will bring, much of which we will cover in this book. There was a consensus that AI would generate economic growth despite the displacement of jobs, but one main concern was how misinformation could threaten democracy. Eileen Donahoe, Executive Director of Global Digital Policy Incubator at Stanford University, listed several human rights concerns, such as equal protection, nondiscrimination, surveillance from AI, and the right to privacy. She also discussed other worries that fall under what she calls social-economic-cultural rights, where the general question is who enjoys the benefits of AI. Furthermore, she brought up the conversation about AI replicating general human intelligence and pointed out that experts can't even agree on what that means: "Are we talking about human consciousness, or human awareness?" she asked. Finally, she raised concerns about the geopolitical implications: "As a foundational technology [AI] has had and will continue to have dramatic implications for economic superiority, military superiority, and really the power to share global norms."[4]

Other powerful and easy-to-use Generative AI tools are also getting much attention. For example, DALL-E 2 is another tool from OpenAI that can generate art and realistic images from a description in natural language. Its power is not only in what it can create but also in its simplicity and ease of use. Anyone with an internet connection, with absolutely no AI knowledge, can create an account, enter a description, and produce artistic images that are astoundingly good. But it doesn't stop there. AI tools are popping up everywhere to create anything from music to presentations. For example, MotionIt AI creates presentation slides, Eightify creates YouTube video summaries, Murf turns text into a human-sounding voice, and Sembly takes meeting recordings and synthesizes them into valuable insights. The list goes on and on.

There are many critics of Generative AI's current limitations. For example, ChatGPT's writing can be bland and uninspiring. But the point is that these tools need not be perfect to be disruptive. Mike Bechtel, Chief Futurist at Deloitte Consulting, shared the perspective that Generative AI may not write better than the best of us, but it does write better than the least of us and arguably than most of us. "The practical standard for any technology isn't supremacy, but rather, utility. To be valuable, a tool doesn't need to be the best tool ever, but simply better than the legacy alternative it sets out to replace," he argues. Despite its limitations, Generative AI has opened the doors to human-AI collaboration at a level never seen before. "We can sit on the sidelines and critique these emerging disruptors, or we can get into the arena and cocreate with them," says Bechtel.[5]

AI is no longer a thing of the future. AI is here now, and our collaboration with machines is just beginning. We have certainly hit an inflection point.

# The Evolution of Human-Machine Collaboration

Since the beginning of human history, our species has leveraged tools to make our lives easier and to accomplish the tasks we need to get done. Over time, these tools became more sophisticated, eventually turning into machines, and the machines became increasingly efficient and smarter. The steam engine brought groundbreaking changes to society, leading to the Industrial Revolution, where human labor became organized to collaborate better with the machines and increase production.

We are now experiencing an even more significant change in how humans collaborate with machines, and the implications are even more dramatic. The difference now is that the intelligence of the machines has increased exponentially, challenging our own intellectual capacity and leading to a whole new set of unknowns as we look into the future. A likely near-term scenario is that we will see an unprecedented productivity increase as these AI-based systems become increasingly sophisticated and as we learn how to best work with them.

We are still in the early days. The experience of working with a tool like ChatGPT can feel like science fiction, and the novelty fuels our fascination. On the other hand, it can cause disappointment due to its limitations. However, this represents an opportunity for humans to bring our intuitive capabilities to the collaborative effort. Current machine-generated works are rehashing other content created elsewhere and may feel bland. They lack soul. They lack the human touch that makes life interesting. That's where we humans come in.

It is becoming clear that AI, automation, and robots will displace many existing jobs. However, at least in the foreseeable future, it appears humans will still have a role to play in production. Those eligible to do the jobs of the future will need new skills that allow them to interface and collaborate with the machines. These skills will constantly change, potentially replacing the four-year college degree with a new paradigm of lifelong learning. A radical new approach to skills development and education will be required. Humans will have to focus on our unique capabilities that can't be replicated by algorithms, while machines perform manual and cognitive work that today requires human labor.

## Superintelligent Soulless Machines

We don't know what the future jobs will be and what exact skills we need to develop. Still, we know that humans have distinct capabilities, such as empathy, creativity, artistic originality, and, arguably, innovative thinking, that are very difficult to replicate through algorithms. Our emotions and intuition are part of who we are. They interact with our bodies and intellect to guide our behaviors, beliefs, and values. A new branch of Artificial Intelligence known as Affective Computing or Emotion AI promises to interpret and replicate our emotions, transforming how we interact with the machines and completely altering our relationships with them. Technology promises to turn our devices not only into personal assistants that understand our every need but also into companions and even intimate friends that know how we feel and respond accordingly.

Even though machines may be able to interpret and possibly even mimic our emotions, reacting to us as if they truly knew us as personal friends, they lack an essential aspect of being human—machines don't have souls. It is crucial to make this distinction because as technology advances and Emotion AI becomes more prevalent, the lines between genuine human connections and interactions with emotionally capable machines will blur. We need to recognize the difference and stay guarded so that our anthropomorphic tendencies—assigning human attributes to inanimate objects—are kept in check. Robots lack free will, so they will never have a genuine affinity toward us. They are simply algorithms created by humans with an almost "magical" ability to detect, interpret, and react to patterns of human emotions and behavior.

This lack of free will allows us to maintain our position as stewards of civilization in a world dominated by machines. The robots may become our servants, colleagues, bosses, friends, or enemies, but they will not do it of their own volition. Machines will always behave according to the capabilities, such as machine learning, and desires, such as objective functions, that we impart to them. It may not always be apparent to us what drives their behavior. AI will perfectly mimic and react to human emotions in ways that can surprise and amaze us. The magnificent processing capabilities of the machines may very well even surpass our intellect. This magical ability will often blur the lines between what is human and what is not.

Much has been written about Artificial General Intelligence (AGI), the potential for machines to achieve general human-level intellect, and Artificial Super Intelligence (ASI), the stage in which machines surpass our intelligence. We have never dealt with an entity that is more intelligent than we are, so we don't know what to expect. We can speculate that AI may become an oracle, a genie, or a sovereign, as Nick Bostrom describes in his book *Superintelligence: Paths, Dangers, Strategies.*[6] These machines may have strategic advantages that give absolute power to potential human sponsors. But unlike humans who have intrinsic motivations, desires, and an inborn need for an identity and a deep understanding of our life's meaning and purpose, the machines will only be able to reflect the motivations we impart to them.

# Human-Machine Interactions

In the future, we will interact with machines in various ways, some even imperceptible. Wearable technologies will become more widespread and integrated into our everyday lives. Some of these technologies may become implantable, and some may even circulate in our bloodstream. Today we benefit from smartwatches that monitor our heartbeat and sleep patterns, among many other functions. In the future, monitoring and reporting vital signs will be pervasive. Blood sugar levels and nutritional deficiencies will be uploaded to our health records synchronously, and any deviations from normal levels will generate alerts or notifications to our healthcare providers. Today we enjoy digital personal assistants, like Alexa and Siri, with near-perfect natural-language understanding. These devices are ready to assist us with questions and can actuate other devices such as light switches, music players, or robot vacuum cleaners. AI experts believe it won't be long before these machines start anticipating our desires, understanding our range of biochemistry and psychosomatic needs, and acting on our behalf without needing our explicit commands.

In August 2020, Elon Musk's company Neuralink demonstrated an implant in the brain of a live pig that wirelessly transmitted detailed brain activity without the need for any external hardware. "It's like a Fitbit in your skull with tiny wires," he said. Musk's goal is to create a Brain-Computer Interface (BCI) that can connect millions of neurons in the human brain and sync our thoughts with Artificial Intelligence machines.[7] His ambition is to take us to an entirely new dimension of interaction with AI to amplify our intelligence. This enhancement may help us maintain parity with the intellectual capabilities of the machines or at least shorten the path between our thoughts and the help we can get from them. It is not hard to imagine a tiny chip implant that would allow us to communicate with others instantly, do a search, or tap into an AI personal assistant that can do all of our planning and scheduling by responding to thought commands. Futurists like Ray Kurzweil have purported other ambitious goals, such as uploading humanity's mental capacities to the "singularity," an entity of sorts with unmatchable intellectual ability that can leverage the power of all of our brains combined. The possibilities are awe-inspiring but also frightful.

We might think that most humans would be reluctant to subject themselves to implants like those developed by Neuralink and other BCI outfits. But we must not forget that intellectual enhancements provide competitive advantages and can be irresistibly enticing. We don't need to look too far to see how performance-enhancing drugs, despite their known health risks, are widely adopted in sports and commonly used by students and professionals in search of extra mental capacity before an important event. Competitive pressures may induce a snowball effect where the more capable the machines become, the more humans will want to directly tap into their intellectual prowess to increase the ability to compete effectively with the machines themselves or with other artificially enhanced humans.

# We Are Unique

Humans have the unique ability to relate to other humans, to feel and demonstrate empathy, to develop intuition, and to be creative in ways that are difficult to duplicate with algorithms. Only time will tell whether machines will be able to reproduce these human skills. In the short term, we have an opportunity to hone our skills in areas out of reach for the machines—the current state of Artificial Narrow Intelligence (ANI) limits the machine's capabilities to performing only very specialized tasks. We can bring a human perspective to a symbiotic relationship with the machines where we do what we are best at, such as empathy, artistic originality, and innovative thinking. The machines can provide complementary capabilities such as search, pattern recognition, and predictive analytics. They can generate text, images, and code from previously created content and perform heavy calculations and data-intensive tasks.

As Affective Computing advances, AI will increasingly become competent personal assistants, colleagues, and general collaborators familiar with our work habits, moods, preferences, and individual idiosyncrasies, knowing us better than any other person. They will therefore be able to optimize how to best work collaboratively with us, helping us increase performance. The way we interact with the machines will become ever more sophisticated. They will be able to not only understand our explicit commands but also anticipate our unspoken needs. They will be able to detect and analyze the subtlest changes in our energy levels, voice intonations, and body expressions, responding accordingly.

This perception ability can be very unsettling. We may not be comfortable collaborating with a machine that knows more about us than we do ourselves. Humans have a unique need for privacy, to keep our thoughts, motivations, and intentions to ourselves. We fear exposing our imperfections and being caught in indiscretions or inappropriate thoughts. Most people would be uncomfortable dealing with an entity that has direct access to our most intimate thoughts and emotions, that knows exactly who we are, imperfections and all. Imposter syndrome is rampant among humans—we have insecurities and certainly would not be comfortable knowing that our collaborators are on to what makes us vulnerable.

This dilemma is a real conundrum that puts us in an uncomfortable position. Will we refuse to collaborate with machines that can detect our feelings and emotions and interpret our thoughts? Who else will have access to such intimate and private information? How do we know the machines' true intent? If we purchase superpowerful personal assistants that can interpret our emotions and perhaps read our thoughts, will they be forever allegiant to us, or will a powerful corporation behind them manipulate us for their benefit? Even if the company that created the machine had the best intentions, what guarantees that a maleficent hacker would not breach the machine's security and gain access to our thoughts and intimacy?

We may eventually find solutions to such problems. Perhaps we will have privacy settings that will limit the machines' capabilities. Still, we will have to make difficult choices that we may not be fully prepared for as we do not fully understand the potential consequences. Settings that limit the machines' access to our emotions, thoughts, and motivations may make them less effective, so there will always be a temptation to trade privacy for convenience or performance. We already make this trade-off today when browsing the Web, searching on Google, and using apps that track more than we sometimes realize.

Anthropomorphism helps us deal with the unfamiliar. Our innate need to interact emotionally and socially helps us connect with others and better understand our world. Our emotional attachment and affection for animals and inanimate objects will be manifested as we interact with robots that will increasingly become more natural and humanlike. As Affective Computing becomes more sophisticated and widespread, robots will interact with humans in ways that may become indistinguishable from our interactions with other humans. It is debatable, but in my view unfeasible, that robots will ever achieve consciousness, especially considering that science doesn't even know with certainty what consciousness is and how it works—more on this later in the book.

But robots will likely perfect detecting, reacting to, and imitating human emotions in ways that evoke real feelings and affections in human beings. We must be cautious that we don't let this become a replacement for genuine human connection and socialization. We might need to create reminders and even restrictions that help us recognize that socially capable robots are nothing more than embodied interfaces we access for our benefit. The misconstruction of anthropomorphic affection for true human love could result in severe psychological and social dysfunctions. Our souls seek genuine human connections that are much deeper than what could be substituted by a robot. The danger is that some people will inevitably be under the illusion that machines can be real friends, confidants, or lovers, preferring to engage with a somewhat predictable entity to avoid the messiness of human relationships while missing the actual benefits of soulful connections.

We have some perplexing ethical dilemmas ahead of us. Technology is advancing much faster than our aptitude to contemplate its consequences and implications, let alone our ability to create effective policies that limit potential dangers. The authorities responsible for creating such policies may not even understand the issues. This lack of clarity is perhaps one of humanity's most significant challenges over the next few decades. A world in which humans and machines collaborate to serve humankind and solve complex problems is a likely scenario. But will we be able to collaborate effectively with machines that continue to evolve with powerful capabilities that leave us baffled with unprecedented ethical dilemmas? Will we be able to make the critical distinction between interacting and socializing with humans—and the benefits these relationships provide to us as individuals and as a society at large—versus interfacing with machines that are very good at mimicking our emotions but lack the gift of the soul?

## Getting Ready

We need to prepare for a future that is approaching at incredible speed. And there is so much preparation required. Our challenge is that we are facing new situations we have never encountered before. Exponential technologies are converging and disrupting industries and legacy institutions. Machines are becoming more intelligent and capable than we are, taking away our jobs and in some cases our identities. The environment has reached its limit due to decades of abuse from our extraction mindset. Educational institutions that insist on hanging on to hundreds of years of old practices cannot keep up with modern requirements. They are resistant to change. Inequality divides us and cries out for justice. The challenges are numerous. And it is all happening much faster than our linear thinking brains and outdated social, economic, and political structures are prepared to handle. We need to upgrade our skills and create new educational programs and institutions that can deliver the preparation required at record pace. We need to quickly embrace containing measures while transitioning to sustainable practices so we don't arrive at an irreversible environmental catastrophe. We need to rethink our socioeconomic paradigms and operating systems. We need healthy debates that will lead to effective policies in light of the many ethical predicaments of technological advances.

Throughout this book, we will discuss many approaches to dealing with these challenges. First, we need to identify uniquely human characteristics that distinguish us from machines that could potentially have more raw intellect than us—machines that can mimic, but not feel, our emotions. By understanding our strengths and weaknesses, we can change our approach to education, focusing our efforts on enhancing the skill sets that will better position us to collaborate with the machines. We could, for example, prepare future generations to improve their creative capabilities, learn the art of innovation, and further develop empathy toward other human beings. We can dedicate time and energy toward these development areas rather than the current emphasis on fact memorization or specialization in domains that the machines will be better able to handle.

To do that, we need to rethink our entire educational paradigm. We must revisit the old methods of subjecting students to information consumption and regurgitation via ineffective testing mechanisms. Using a college degree as a proxy for measuring people's suitability for jobs must be reconsidered. The current economic model of higher education that requires students to start their adult life acquiring life-altering debt to pay for a college degree must be disrupted. The idea that one should concentrate their entire higher education pursuit in a single four- to five-year concentration must be reexamined. The centralized credentialing mechanism in which a university confers the all-powerful college degree must be replaced by a technology-based decentralized credentialing process. Students must be able to accumulate many educational credits from multiple institutions of higher learning over an extended period, leading to certificates, nanodegrees, and complete degrees.

With the help of technologies such as AI and Virtual Reality (VR), personalized learning will completely alter how we think about education. AI will be able to customize how we access and retain knowledge based on our learning preferences. It will be able to detect whether we are engaging with the material and learning from it and will make changes accordingly. Material and delivery methods will then be optimized to our individual learning styles based on whether we respond best to visual, auditory, experiential, or other learning forms. VR will allow us to immerse ourselves in a universe of rich experiences that tap into our emotions and intellectual interests to enhance skill development. This individualized approach will include priming the brain and body to become more receptive and retentive. Advanced interfaces that allow us to search, retrieve, and store information in unimaginable ways may completely transform the way we learn, opening doors to yet inconceivable possibilities.

Despite all the incredible opportunities a new education paradigm promises to enable, we can only realize them if we tap into our will, emotions, and motivations. We need to understand what drives and inspires us. We must recognize the inner forces that contribute to our actions and make us who we are.

The challenge with any discussion involving human will, emotions, and motivation is that, despite many advances in science, there is still so much we don't understand. Is it possible that the complexity of what it means to be human was not meant to be understood by our limited intellectual capabilities? One of the great mysteries of our existence is that we know that we are conscious beings, yet we struggle to define precisely what consciousness is, as we will see in a later chapter.

Psychologists refer to the Theory of Mind as a social-cognitive skill that we learn as children allowing us to think about our mental states and those of others. It is the human ability to think about thinking. It is a fundamental aptitude for socializing, understanding how others think, predicting their behavior, and solving interpersonal conflict. Theory of Mind helps us create a frame of reference to recognize and ascribe mental states to others that are separate and distinct from our own, yet it does not explain consciousness. The big-picture aspect of consciousness baffles scientists and remains undefinable. We intuitively know that it relates to awareness, realization, learning, and meaning, but this is not enough to translate it into something we could build into a machine. Perhaps consciousness is a matter of the soul that will forever remain a mystery.

Despite our struggles with many aspects of human existence that we don't fully understand, we have learned a considerable amount about human behavior, motivations, and performance in the last few decades.

Technologies like fMRI (Functional Magnetic Resonance Imaging) have given us new insights into how our brains function. Discoveries such as brain plasticity, brain priming, and the effect of physical activities and breathing in a meditative state have opened new doors for improved learning and performance. We have learned, for instance, that our motivations can be influenced much more effectively with methods and techniques that transcend the old paradigm of carrots and sticks.

Extrinsic rewards and punishments can backfire in surprising ways. Instinctively, we might think that giving people external rewards such as cash or bonuses might improve their performance. But that is not always the case. External rewards work well for tasks that require following a set of instructions. But it is a terrible motivator for tasks that require creativity and innovative thinking—for these types of tasks, the external reward may actually lower performance.[8] Human performance is influenced not only by external factors but also by intrinsic motivators. Several elements, including autonomy, mastery, and purpose, induce intrinsic motivation. Intrinsically motivated work feels more like play, but if you introduce an extrinsic motivator, like a reward, the play turns back into work. The unintended consequence of such extrinsic reward is a reduced ability to fully engage our capacities for innovative and creative thinking, negatively impacting performance.

This phenomenon is particularly relevant as we consider our engagement with machines. Machines are becoming increasingly capable of performing tasks described in a set of instructions, the type of tasks known as algorithmic for which extrinsic rewards work well. Where humans can excel and bring the most value to a collaborative engagement with machines is in performing intrinsically motivated tasks. Humans will have an important role, at least in the foreseeable future, in addressing what some authors refer to as "wicked problems." Wicked problems are problems that don't have any right or wrong answers. These problems cannot be solved by applying a formula or using prescribed methods that have worked in the past. They feel more like chasing a moving target. These are problems that designers deal with all the time. They require innovative, creative, artistic abilities that may involve a sense of aesthetic or poetic talents. Humans have an affinity for beauty, for awe-inspiring works that seem to touch a soft spot in our souls—something that the machines lack.

Humans will be able to truly contribute to a symbiotic relationship with machines by using our intuitive ability to connect the dots in unique ways. We may not be able to accomplish the specialized tasks that the machines will do incredibly well. Still, we will be able to contribute as aggregators, synthesizing the output from the machines to create something new and unique that requires a human touch. To effectively perform these types of tasks, we must focus on becoming lateral thinkers. Lateral thinkers can search several domains to find solutions to apply to other domains. They are polymaths, knowledgeable in many different areas, capable of leveraging empathy to understand multiple points of view, and incredibly skilled in asking the right questions to elicit answers that will generate new insights. The ability to ask the right questions will be one of the most valuable human skills in the future.

Neuroscience has given us new insights into why some people can work endless hours and feel a sense of joy and purpose while many workers find themselves disengaged, overstressed, or bored. We now understand that optimal performance happens when the stress level is just enough to stretch our abilities beyond our comfort level in a voluntary effort to accomplish something difficult or worthwhile, but not so much that it causes burnout. Psychologist Mihaly Csikszentmihalyi, who is considered the father of flow, in his initial research reveals that his subjects frequently called the flow state "addictive" and admitted to going to exceptional lengths to get another "fix." In his book *Flow: The Psychology of Optimal Experience*, he states, "The [experience] lifts the course of life to another level . . . Alienation gives way to involvement, enjoyment replaces boredom, helplessness turns into a feeling of control . . . When experience is intrinsically rewarding life is justified."[9]

Once we find ourselves in a state of flow, the intrinsically rewarding nature of the experience compels us to continue working regardless of any extrinsic rewards associated with the work. In other words, the work itself is its reward. Csikszentmihalyi explains it further:

*In a culture supposedly ruled by the pursuit of money, power, prestige, and pleasure, it is surprising to find certain people who sacrifice all those goals for no apparent reason. By finding out why they are willing to give up material rewards for the elusive experience of performing enjoyable acts, we learn something that will allow us to make everyday life more meaningful.*

The science behind flow is compelling and will be fundamental to our well-being as we enter a world potentially marked by abundance and joblessness. Abundance may result from the extraordinary technological advances that will drive sustainable productivity, creating an unprecedented level of material goods and promising to address the basic needs of humanity. Joblessness may result from machines doing most of the work required to produce this abundance. Humanity will have to choose between turning into zombies, bored, addicted, grabbing on to vices in search of meaning and purpose, or turning into a more introspective and connected species that care for one another. The second choice means dedicating time and energy toward justice, artistic expression, and spirituality, fulfilling the unique human need to fill a void in our souls—we will discuss these topics in more detail throughout this book.

But before we can get to a state of abundance, we will have some challenges to overcome.

## Crossing the Chasm

The exponential changes driven by technology will be deflationary in the long term. Therefore, the price of goods and services should continuously decline, in many cases approaching free, driving the demonetization of the economy. In a world where technology-driven productivity creates abundance, the standard of living would increase, and basic human needs like food and shelter might become available at negligible cost. In this scenario, work might become unnecessary, and we could spend our time focusing on matters of the soul, creating a revival of artistic expressions and spirituality. Machines can do the soulless work, while humanity reaches for higher purposes.

But before we get to this utopian scenario, where machines do all the work and produce abundance so that humans can dedicate more time and energy to the needs of the soul, we need to cross the chasm that separates it from our current reality. In the early days of human history, our primary preoccupation was survival. Our ability to communicate and cooperate allowed us to survive and multiply as a species. This multiplication meant that we had to continuously share the available resources, given the limitations imposed by our capabilities and geographical reach at any time. These limitations resulted in the scarcity mindset that has defined us to this day. As the number of people continued to grow, the resources became scarce, forcing us to expand our capabilities to extract resources more efficiently and explore new geographies where resources were more abundant.

This scarcity mindset resulted in an operating system based on the foundational principles of extraction, exploitation, and hoarding of scarce resources. To continue multiplying, humans had to find ways to stimulate growth at any cost. We pushed this growth imperative to its limit until we reached the point where unrelenting extraction of our resource-limited Earth became unsustainable. We are only now realizing that perhaps we have taken this too far. The environmental price that we have paid is more than we can afford. Deforestation, the decimation of a variety of species, pollution, greenhouse effects, and other environmental harms are starting to reveal their consequences in the form of global warming, rising sea levels, forest fires, destructive hurricanes, respiratory diseases, and more.

The growth imperative has resulted in a socioeconomic system that is also unsustainable. Modern monetary policies have sustained growth by issuing debt, printing money, and manipulating currencies. According to Jeff Booth, author of *The Price of Tomorrow*, over the last twenty years, we have issued $185 trillion of debt to produce about $46 trillion in GDP growth. This debt stimulus has artificially maintained central

banks' mandates to keep unemployment low. It has also prevented deflation, despite the deflationary forces of exponential technologies. Booth
explains the problem: "The only way to keep our economies growing
and combat the effect of that exponential technology under the existing
system is to allow debt to rise exponentially as well."[10] Governments insist
on issuing additional debt, kicking the can down the road. Eventually,
the problem becomes so big that the only solution becomes pressing
the reset button and restructuring this enormous debt, which could
result in more instability than if we had just let the deflationary forces
run their course.

This debt-laden growth stimulus has also resulted in another unintended consequence: inequality. Inflation erodes the buying power of
cash while increasing the price of assets. Consequently, in an inflationary environment, the wealthy are rewarded by the increasing value of
those assets. Workers, on the other hand, depend on wages that, for the
most part, have stayed stagnant for the last few decades while the price
of goods continues to rise. The owners of assets have benefitted enormously from this artificially created growth resulting in an unsustainable
increase in inequality. According to the Pew Research Center, "the wealth
gap between America's richest and poorer families more than doubled
from 1989 to 2016. In 1989, the richest 5% of families had 114 times as
much wealth as families in the second quintile, $2.3 million compared
with $20,300. By 2016, this ratio had increased to 248."[11] According to
Oxfam International, the world's richest 1 percent own twice as much
as the bottom 90 percent,[12] This blatant inequality can give rise to social
unrest, extremism, political upheaval, and possibly wars

So how do we cross the chasm? How do we transition from the current
dystopian reality of environmental degradation, inequality, social unrest,
and injustice? How do we reach abundance, sustainability, and freedom
from meaningless labor so humans can put their time and energy toward
seeking the higher purposes of the soul? This walk will not be easy, but we
must find our way. We must come together, respectfully debate alternatives, achieve compromises where needed, provide safety nets for those
most vulnerable, and continue our march toward a shared vision and goal

of a better world. We must ensure every human has their basic needs met and that our only home, Earth, can be preserved for many generations.

## Taking Care of Our Souls

"How's your soul?" This is how a letter by Pastor Joel Johnson from Westwood Community Church in Chanhassen, Minnesota, began as he addressed his congregation. In the letter, he states:

> *Everything that matters most flows from the inside out. Soul care opens the door to God's presence. He is, and always will be, the best way to find meaning and adventure, peace and fulfillment. I pray your soul is well.*

Regardless of your religious convictions, the idea of taking care of the soul should resonate. There is something uniquely human that is incredibly difficult to describe but easy to relate to and understand—this mysterious aspect of the human condition called the soul. The soul is much more than just consciousness or emotions. It is a part of who we are that transcends all organic and material matters. We could have everything going for us—health, finances, relationships, and social status—but if our soul is not well, then we are not at peace. Surprisingly for many people, the opposite can also be true. Even if we find ourselves in very adverse circumstances in all aspects of our lives, we can still find peace if we can address the matters of the soul.

This is where the machines can't reach. It is our true differentiator. It is the one thing that can't be simulated, mimicked, translated, digitized, replaced, outsourced, automated, or taken away from us. In a world surrounded by machines, we must focus on taking care of our souls.

We are entering a time unlike any other in human history when change is happening at unprecedented speeds. We must make critical choices that could lead to a life of peace, abundance, and fulfillment, or result in conflict, inequality, and dissatisfaction. I hope this book will help bring light to the opportunities and challenges ahead so that we can engage in constructive dialogues, vote for those worthy of representing us, and make choices that are well informed and that lead to a fulfilling life for many generations to come.

Let's begin our discussion by reviewing humankind's ultimate invention. This game-changing, nothing-will-ever-be-the-same-again technology that will completely redefine our work and challenge our superior intellect: Artificial Intelligence.

## Takeaways

» The advent of ChatGPT has popularized Artificial Intelligence, which could represent an inflection point in the growth of AI.
» This introductory chapter presents vital concepts to be discussed throughout this book, including:
  • Our interactions with the machines will change;
  • We are different from machines despite AI's advancing ability to mimic human emotions;
  • We need to get ready for a vastly different environment surrounded by AI;
  • We must cross the chasm between our reality and a possible bright future where we will have more time and energy to address the needs of the soul.

CHAPTER 2:

# Humankind's Ultimate Invention

Several technology platforms with exponential growth have surfaced in the last few decades including Artificial Intelligence, Biotechnology, Nanotechnology, Robotics, 3D Printing, Blockchain, Augmented and Virtual Reality, and others. This is the first time we have seen so many technology platforms with such astonishing transformational power come together so rapidly. The last time this occurred was in the late 1800s and early 1900s when only three such platforms—electricity, telephony, and the internal combustion engine—came into the world to completely transform society and create the foundation for our modern era.

Imagine the future impact when dozens of exponential technology platforms start to converge, as we are experiencing today. According to Ray Kurzweil, the result is that we will experience twenty thousand years of technological change over the next one hundred years. As incredible and transformative as all of these platforms are, one stands out: Artificial Intelligence.

Some authors, like James Barrat, believe that AI will be "man's last invention." In his book, *Our Final Invention: Artificial Intelligence and the End of the Human Era*, he discusses the potential risks and benefits of human-level or superhuman Artificial Intelligence.[13] Those supposed risks include the extermination of the human race. Others, like Nick Bostrom, suggest that we may be able to control a superintelligent agent so that it will behave in ways that are beneficial and not harmful to humans. In *Superintelligence: Paths, Dangers, Strategies*, he describes several control methods, all of which have their own potential benefits and shortcomings.

Will AI become our last invention due to our inability to control it, resulting in humanity's extermination? No one knows, so this is highly speculative. However, evidence continues to indicate that AI may become so powerful that it will be able to create new inventions beyond what humans are capable of developing. This ability became abundantly clear when Google's DeepMind algorithm AlphaGo Zero learned on its own to play the games of chess, shogi, and Go, defeating the best computers that had been fed instructions from human experts. AlphaGo Zero, by itself, came up with playing strategies that humans had never thought of before.

But it is not just in games that AI is surpassing human capabilities. AI is being implemented to solve real-world problems, leading to exciting discoveries. In 2020, AI solved a fifty-year-old highly complex challenge in biology called protein folding, which will be critical to many future medical advances. With the help of AI, humans will unveil new possibilities beyond our current limited reach. The moniker "humankind's ultimate invention" is perhaps more fitting for the machine that can invent and create everything that we need beyond our wildest imagination.

In an interview with Karen Hao, Geoff Hinton—one of the pioneers of deep learning—compared Artificial Intelligence to the capabilities of the human brain: "I do believe deep learning is going to be able to do everything, but I think there are going to have to be quite a few conceptual breakthroughs. The human brain has about 100 trillion parameters, or synapses, whereas what we now call a really big model, GPT-3, for example, has 175 billion. It's a thousand times smaller than the brain." Hinton is alluding to the breakthroughs required to achieve what is referred to as Artificial General Intelligence (AGI), the human-level intelligence after which the original Artificial Intelligence term was coined. In early 2023, GPT-4 was rumored to contain the 100 trillion parameters equivalent to the human brain but Sam Altman, CEO of OpenAI, later denied it.[14] But eventually, GPT-n may get us there. GPT-4 was released just before this book was finished. It is based on a Multimodal Large Language Model, which means that it uses images, video, audio, and other sensory data besides text. In other words, it is becoming more perceptive, driving the speculation that it could get us closer to AGI.[15]

In the historical progression of Artificial Intelligence, computational power has always been a prerequisite to the next level of progress. Getting to AGI is no different. Scientists have estimated that the human brain is capable of approximately ten quadrillion computations per second (cps). The world's fastest computer has already beaten this number but at an extraordinary cost. Kurzweil, who is very optimistic about Artificial Intelligence, estimates that AGI can become widespread when we get ten quadrillion cps for about $1,000. Using Moore's Law, which has predicted the growth of computing power so far, it is estimated that we will be there by 2025. But getting to AGI requires much more than just raw computational power. Kai-Fu Lee, a distinguished AI expert, does not believe we will achieve AGI anytime soon: "There are many challenges that we have not made much progress on or even understood, such as how to model creativity, strategic thinking, reasoning, counterfactual thinking, emotions, and consciousness. These challenges are likely to require a dozen more breakthroughs like deep learning, but we've had only one great breakthrough in over sixty years, so I believe we are unlikely to see a dozen in twenty years."[16] Despite tremendous progress with the advancement of machine learning algorithms, AGI continues to be an elusive goal.

*Exponential View*, a technology newsletter, portrays the point of view of Gary Marcus, the CEO and Founder of Robust.ai, who is also a psychologist and neuroscience professor:

> *I do think that we'll be surprised by what we'll do with deep learning. The test will come when GPT-n has more*

*parameters than the brain, which may be only five to ten years away. Will a 100-trillion-parameter neural network outperform a 100-trillion-connection brain? I doubt it. Will it be an amazing piece of technology? Certainly. What about a 10-quadrillion-connection neural network? What will that be like? And what will we call the things it can do?*

We can debate if and when we will achieve AGI or even Artificial Super Intelligence. But there is no denying that AI, even in its current state of Artificial Narrow Intelligence, will bring about socioeconomic changes of unprecedented magnitude.

# The Evolution of AI

As we look at the future of AI, it is helpful to look back to understand how it evolved. Three growth vectors have impacted the evolution of AI: algorithmic advances, computing power, and data explosion. These vectors have their historical landmarks as outlined below until they converged around 2007. In the last decade, AI has taken a revolutionary leap forward, much of it driven by deep learning, a machine learning technique based on multilayered artificial neural networks. But before diving into deep learning and its practical applications, let's briefly review the history of these three critical growth vectors.

The first growth vector, algorithmic advances, goes back as far as 1805 when French mathematician Adrien-Marie Legendre published the least square method of regression, which provides the basis for many of today's machine learning models. The architecture for machine deep learning using artificial neural networks was first developed in 1965. Between 1986 and 1998, we saw several algorithmic advances: backpropagation, which allows for optimization without human intervention; image recognition; and natural language processing, to name a few.

The second growth vector, computing power, had a significant historical landmark in 1965 when Intel's Cofounder Gordon Moore recognized the exponential growth in chip power with the observation that the number of transistors per square inch doubles every eighteen months. This observation became known as Moore's Law and correctly predicted the growth of computing power to the present day. At the time, the state-of-the-art computer was capable of processing in the order of three million FLOPS (floating-point operations per second). By 1997, IBM's Deep Blue achieved eleven billion FLOPS, which led to its victory over Gary Kasparov, the world chess champion. In 1999, the Graphics Processing Unit (GPU) was unveiled—a fundamental computing capability for deep learning. The GPU has rapidly evolved into the primary hardware platform for AI. In 2004, Google launched MapReduce, which allows computers to deal with immense amounts of data by using parallel processing. Next, deep learning advancements led to the emergence of neuromorphic chip designs that instantiate neurons directly in silicon. According to one analysis by Gartner, neuromorphic designs will vastly displace GPUs as the primary hardware platform by 2025.

Finally, the third growth vector, data explosion, started in 1991 when the World Wide Web was made available to the public. In the early 2000s we saw wide adoption of broadband, which opened the door to many internet innovations, resulting in Facebook's debut in 2004 and YouTube in 2005. At that time, the number of internet users worldwide surpassed one billion.

The year 2007 became a significant landmark. At this point, the technologies began converging as the mobile explosion came to life with Steve Jobs' announcement of the iPhone. By 2010, three hundred million smartphones were sold, and internet traffic reached twenty exabytes (twenty billion gigabytes) per month.

From there, several significant advances gave birth to a renewed enthusiasm for Artificial Intelligence. In 2011, IBM Watson defeated the two greatest Jeopardy! champions, Brad Ruttner and Ken Jennings. Such achievement was made possible by IBM servers capable of processing eighty trillion FLOPS. Remember that when Moore's Law was pronounced in the mid-1960s, the most powerful computer could only process three million FLOPS, so at this point computing capacity had increased by more than twenty-five million times in just over four decades.

The year 2012 marked a significant milestone in AI's history with the unveiling of deep learning, a breakthrough that has fueled many practical AI applications in use today. In an annual event called the ImageNet Large Scale Visual Recognition Challenge, teams from the world's leading universities and corporations compete to identify images from an extensive photograph database. Geoff Hinton and his team from the University of Toronto's research lab trained their algorithms using a multilayered Convolutional Neural Network (CNN/ConvNet) that vastly outperformed other algorithms. This achievement created new excitement in the AI research community, which started to develop new capabilities by coupling massive datasets with these neural algorithms. This advancement became possible due to the enormous number of images and videos captured by smartphones and shared on social media at the time. This data abundance coincided with the affordability of large storage and computing power, creating a fertile environment for this type of research. Google used sixteen thousand processors to train a deep artificial neural network to recognize cat images in YouTube videos without providing any information about the images to these machines. CNN became capable of classifying images with a high degree of accuracy, producing results that seemed like science fiction just a few years back. In the meantime, the data explosion continued, with the number of mobile devices on the planet exceeding the number of humans.

In 2015, Google released TensorFlow, a comprehensive software platform for deep learning. This release has allowed researchers and engineers to develop practical applications, optimize code, and build new tools that allowed them to take a higher-level perspective as they built new AI systems. In 2017, DeepMind's AlphaGo defeated Ke Jie, then the world's top-ranked Go player, creating another wave of excitement and wonder across the AI research community. Go is a very complex game that draws heavily on human intuition, or what the best players describe as simply a "feeling" that guides their judgment as they move the pieces across the board. This type of capability seemed to be beyond the reach of computers. At this point the general belief that certain types of jobs were safe from the threat of automation started to crumble.

The technique used by the DeepMind team to defeat the world's top-ranked Go player is generally known as reinforcement learning. Initially, supervised learning is used to feed millions of moves from the best human players to the AI. Once the AI acquires this initial knowledge, it is let loose to play against itself. An objective function drives the AI to improve with each round, resulting in gradual progress toward superhuman—albeit narrow—capabilities. The achievement of AlphaGo was so influential that it prompted the Chinese government to set its sights on becoming the undisputed leader in Artificial Intelligence, an event described as a "Sputnik moment" by Kai-Fu Lee.

This AI research breakthrough also resulted in renewed hopes for achieving AGI, and the relentless search continues. In 2017, researchers at Google created a technology called "transformer," a model capable of exhibiting selective memory that remembers anything important and relative in the past. This capability allows AI to contextualize text and teach itself a new language using self-created constructs and abstractions rather than human constructs like conjugation and grammar. Google's transformer work has been extended in collaboration with other AI researchers, resulting in what is known as Generative Pre-trained Transformer, or GPT, the technology behind the now popular ChatGPT.

OpenAI, an AI research and development organization founded by tech luminaries like Elon Musk and backed by Microsoft and others, has developed additional breakthroughs that continue to feed the race toward AGI. In 2019, OpenAI unveiled GPT-2 using a generative neural network trained on massive troves of text downloaded from the internet. A generative neural network system is optimized to create samples similar to its training data. This optimization uses a technique called Generative Adversarial Network (GAN). In essence, the AI "plays" against itself millions of times, just like it did as it learned the game of Go, with an objective function set to minimize errors, or differences, between the generated samples it creates and the original data fed to it.

GAN is also the technology behind what is known as deepfakes, fabricated digital creations that are difficult to distinguish from originals, such as fake videos of celebrities or political figures that contain the same voice, intonation, and mannerisms as the actual person. GAN is so effective that the differences between real and fake videos are totally imperceptible to humans. Deepfake is of great concern as it can potentially be used to disseminate false information. However, the technology that allowed deepfakes to proliferate can also detect and eliminate them. Anti-deepfake software will likely be used as widely as antivirus software.

AI's evolution continues to unfold at a rapid pace. In 2020 OpenAI released GPT-3, increasing the number of parameters it uses a hundredfold to 175 billion. Using one of the most powerful computers in the world, GPT-3 was trained on more than forty-five terabytes (forty-five thousand gigabytes) of text and is growing tenfold each year. With all the knowledge accumulated from this massive amount of human-created text, GPT-3 can now write poems, press releases, and technical manuals. It can also simulate various writing styles. Despite these fantastic advancements, it is vital to remember that GPT-3 still lacks causal reasoning, abstract thinking, and common sense, so we are still nowhere near arriving at AGI. As we progress, we must be guarded against imparting human biases, prejudice, and malice into AI, as already evidenced in GPT-3. We will further discuss ethical concerns at the end of the chapter. As previously mentioned, GPT-4 was released just before this book was completed, but OpenAI did not divulge technical details such as the number of parameters it uses.

As discussed earlier, computational power is a critical ingredient in the search for AGI. Intel has unleashed a cloud-based experimental neuromorphic computing system containing one hundred million hardware neurons—roughly the equivalent of a small mammal's brain. But experts believe that we need a new computing paradigm to achieve the level of computing capability that will lead to AGI. This is where quantum computing comes in.

As astounding as it is, everything in traditional computers, from complex calculations to high-resolution images and videos to extensive communication networks, is based on the fundamental concept of a bit, which is simply a switch that is either off, representing a zero, or on, representing a one. By combining zeros and ones, we have developed marvelous devices that intersect every aspect of modern civilization. However, we are starting to hit the physical limits of our ability to pack more and more bits into tiny chips measured in nanometers. As is often the case with breakthrough innovation—to overcome what seems to be an insurmountable physical limitation—a new paradigm was invented called quantum computers.

Instead of using bits, which are limited to the binary states of zero and one, quantum computers use qubits that follow quantum mechanics principles and use the properties of subatomic particles' behavior. One such property is superposition, which allows the qubit to be in multiple states at any given time. A combination of multiple qubits in superposition enables quantum computers to process a vast number of outcomes simultaneously, and that is what gives these computers super-processing capabilities.

However, these super-processing capabilities come at a price. Quantum computers are susceptible to small disturbances such as vibrations, electric interferences, temperature changes, and other environmental factors. Quantum computers are kept at extremely low temperatures to overcome these restrictions. This limitation makes it challenging to scale, but even with just a few qubits, quantum computers can process some tasks over a million times faster than traditional computers. In 2019, Google demonstrated a 54-qubit quantum computer that could solve in minutes what would have taken a conventional computer several years. IBM expects to double the number of qubits yearly for the next few years and is planning to release a 1,000-qubit processor in 2023. There are great expectations that we will witness significant progress in the next five to ten years. In twenty years, quantum computers are expected to be so powerful that the state-of-the-art encryption used today to safeguard cryptocurrencies will be at risk of being cracked, creating a new set of potential disruptions.

Quantum computers will require the development of new algorithms and software tools to become widespread and useful on a massive scale. This software will likely be developed in the form of open-source.

High-tech companies like Google, Facebook, and Baidu are releasing much of their software in open-source form. The most advanced research conducted by organizations like DeepMind and OpenAI is published openly in leading scientific journals. However, it is important to point out that we do not know to what extent proprietary information is not being shared.

The accessibility of cloud-based quantum computing combined with freely available advanced research in the form of open-source software will unleash a new stage in AI's evolution. The impact these advances will have is unlike anything we have seen so far, opening the door to breakthroughs that can potentially lead to AGI.

However, deep learning advancements require an enormous amount of data, an inhibitor limiting AI's dominance to large tech companies or government entities that own the data. Case in point, several companies are vying for leadership positions in developing autonomous vehicles. Given the complex and diverse driving conditions one may encounter, the AI guiding these vehicles must be trained with an enormous amount of data. To address this need, Tesla has equipped every vehicle with eight cameras that operate continuously. Data captured by these cameras is selectively uploaded to Tesla's data center, continually adding to Tesla's massive trove of real-world visual data used to train its vehicles, providing an unmatched competitive advantage.

Google has access to a colossal amount of text data captured via its search engine, Gmail, and other applications. Using this data to train AI, Google can produce text and complete sentences for its users in many languages. It can also instantly translate languages with an astonishing degree of accuracy. Amazon's Alexa is listening to conversations in homes across the world. This data is essential to train AI in understanding customer behavior, but it raises several ethical questions, as we will see next.

# Ethical Concerns

AI's increasing influence is causing grave ethical concerns as we grapple with how to set boundaries for this technology of unparalleled capability.

According to political philosopher Michael Sandel, "AI presents three major areas of ethical concern for society: privacy and surveillance, bias and discrimination, and perhaps the deepest, most difficult philosophical question of the era, the role of human judgment." If AI can outsmart us, should we leave it up to AI itself to make critical ethical decisions for us, or, as Sandel asks, "Are certain elements of human judgment indispensable in deciding some of the most important things in life?"

Let's consider each of the ethical concerns raised by Sandel.

## PRIVACY AND SURVEILLANCE

Digitization and advances in AI have brought many societal benefits and will continue to do so. However, the price we are paying is the deterioration of privacy. Every time we interact with a digital device, we leave breadcrumbs that track where we have been, what we have seen, and what we have said. Our Web browser cookies track the websites we have visited. Email and texting apps track our communications. Global Positioning Systems (GPS) on our mobile phones track our location. Surveillance cameras everywhere contain images that can be traced back to an individual through face-recognition technologies. Even in the sanctity of our homes, Alexa, Siri, and other personal assistants are listening to the words we say, whether they are meant to remain private or not. And this is just the beginning.

During the pandemic, we have seen a dramatic increase in home office work, which has driven the demand for remote surveillance technologies. Employers concerned about worker productivity are deploying these technologies to monitor how employees spend their time away from the office and what they are doing when not being watched by hall-walking supervisors. Precision economy is a term being used to describe a future workplace of hyper-surveillance and algorithmic optimization that measure and incentivize virtually every aspect of our working lives. New advances in neurotech—technology that combines neuroscience and AI to achieve brain surveillance—allow employers to monitor the brains of employees, raising the level of concern to unprecedented levels.[17]

Employee monitoring is not a new practice. What is disturbing in this new age of technological surveillance is that we are using these tools to access employee behaviors in the privacy of their homes, where audio and video surveillance can capture private spaces and conversations, raising the level of intrusiveness. Other concerns are the use of these technologies to possibly harass or intimidate workers. AI and other analytics tools used to assess performance may lead to biased treatment of employees, discrimination, and lower morale.[18]

A stern concern regarding AI's invasion of privacy is that many consumer products, from smart home appliances to smartphone apps, exploit data without the user's awareness. Most users take a casual approach to privacy and provide permission for data exploitation without fully understanding the extent of those permissions. No one has the time or patience to read several pages of privacy policy documents, so disclosures are highly ineffective when providing consumer protection.

Other efforts to protect privacy include the anonymization of data collected. However, with AI's ability to process and analyze vast data quantities, it can de-anonymize data on inferences from multiple devices that track and monitor individuals. Voice and facial recognition capabilities that can severely compromise anonymity in the public sphere are particularly concerning. For instance, law enforcement can now use voice and facial recognition to find individuals without probable cause or reasonable suspicion, thus circumventing legal procedures.

AI's predictive analytics capabilities can infer or predict sensitive information from nonsensitive data sources. For example, keyboard typing patterns, facial hues, voice characteristics, blinking frequency, and body temperature can be used to deduce many emotional states, from nervousness to confidence and sadness. Other metrics and data logs can be used to identify a person's political views, ethnic identity, sexual orientation, and overall health. This information can be used to sort, classify, evaluate, and rank people without explicit consent. China's social scoring system is a clear example of how these technologies can impact a person's creditworthiness, employment, and access to housing or social services.[19]

The emerging field of privacy computing is researching ideas to address concerns regarding preserving privacy. One example is federated learning, a technique in which AI is trained across multiple decentralized servers holding local private data without direct access. The AI approximates centralized learning while not accessing the local data. Another idea is homomorphic encryption, where AI is trained on encrypted data. Whether these technologies will effectively protect our privacy is yet to be seen. For now, data privacy regulations are the fundamental mechanisms used to minimize privacy evasion's harmful effects.

## BIAS AND DISCRIMINATION

AI can develop biases based on the data it is trained on. We have seen the data explosion resulting from the advent of the internet and social media. This rich data source is widely used to train AI models, and herein lies the danger. Much of the data in the online world comes from anonymous human contributors, implying that AI will pick up on the worst characteristics of human beings because of the anonymity of the data source.

A new study by the Chinese Academy of Science (CAS) reveals that many chatbots show depression and addiction symptoms when asked questions generally used as a cursory intake for depression and alcoholism. The study found that all the bots surveyed, including Facebook's Blenderbot, Microsoft's DialoGPT, WeChat, Tencent's DialogFlow, and Baidu's Plato, scored very low on the empathy scale. CAS's Institute of Computing Technology became curious about the mental health of these bots after reports emerged in 2020 about a medical chatbot telling a test patient that he should kill himself.

The Chinese researchers found that all the assessed chatbots exhibited several mental health issues. In reality, the AI does not "suffer" from these human ailments as they can't feel anything, but they do develop biases based on the data characteristics they have been trained on. The study shows the potential source of the problem: All these chatbots from Facebook, Microsoft, Baidu, and WeChat/Tencent were pre-trained using comments from Reddit, a social media platform known for its negative commentary.[20]

Dr. Sanjiv M. Narayan, professor of medicine at Stanford University School of Medicine, provides a perspective on how biases arise in AI: "There is an increasing focus on bias in Artificial Intelligence, and while there is no cause for panic yet, some concern is reasonable. AI is embedded in systems from wall to wall these days, and if these systems are biased, then so are their results. This may benefit us, harm us, or benefit someone else."

He then elaborates: "A major issue is that bias is rarely obvious. Think about your results from a search engine 'tuned to your preferences.' We already are conditioned to expect that this will differ from somebody else's search on the same topic using the same search engine. But are these searches really tuned to our preferences or to someone else's preferences, such as a vendor's? The same applies across all systems."

The challenge with preventing bias from affecting AI's usability is that bias can be infused in many different ways. "Bias in AI occurs when results cannot be generalized widely. We often think of bias resulting from preferences or exclusions in training data, but bias can also be introduced by how data is obtained, how algorithms are designed, and how AI outputs are interpreted," he says.

According to Dr. Narayan, multiple techniques exist for keeping bias out of AI, but none are foolproof. He believes that the first step is understanding the various causes of bias. "The technology of AI is moving inexorably toward greater integration across all aspects of life. As this happens, bias is more likely to occur through the compounding of complex systems but also, paradoxically, less easy to identify and prevent," he says.[21]

We should expect AI to continue showing biases if it is to function in the complex world of human activities. Just like a child picks up on the culture, beliefs, and behaviors of parents, AI will learn whatever biases are embedded in the data it is trained on. It is unreasonable to expect any data source containing human interactions and communications to be free from biases. Unless we want to exclude AI from operating in the world of human activities, this will continue to be a challenge.

## HUMAN JUDGMENT

What is the role of human judgment in a world where AI's capabilities are becoming ever more pervasive, impacting every aspect of our lives, including life-or-death decisions? Behind this simple question, there are several complex considerations.

It starts with understanding human judgment. Humans make judgments that are influenced by several factors, including values, culture, experiences, emotions, and intellect. We recognize that our judgments can be biased, unjust, and sometimes outright evil. Given these shortcomings, to what extent should we impart our judgments to AI? Considering that AI may someday surpass our intellectual abilities, should humans be seeding the values that will drive AI's judgment? Would it be wiser to leave it up to AI, with its potentially superior intellectual capabilities, to determine its own values driving its judgment?

Eliezer Yudkowsky, an American researcher popularizing the idea of a friendly AI, has proposed seeding the AI with what he calls our coherent extrapolation volition (CEV). Here is how Yudkowsky defines CEV:

> *Our coherent extrapolation volition is our wish if we knew more, thought faster, were more the people we wish we were, had grown up further together; where the extrapolation converges rather than diverges, where our wishes cohere rather than interfere; extrapolated as we wish that extrapolated, interpreted as we wish that interpreted.*

What Yudkowsky is trying to do with CEV is come up with a morality model that encapsulates moral growth but keeps humanity ultimately in charge of its destiny. Humanity must grapple with this fundamental question: What moral values do we want to seed AI with? If we don't know, are we comfortable letting an all-powerful algorithm determine the moral values that will guide its judgment without human interference?

These challenging questions will become ever more pressing as AI proliferates in all aspects of our lives. We will increasingly be working, collaborating, and living with these intelligent agents that may soon match or even surpass at least some aspects of our intellectual capabilities. Our ability to effectively deal with this unprecedented scenario will be paramount to the future of humanity.

In the next chapter, we will consider how humans may collaborate with these intelligent machines.

## Takeaways

» AI can be considered humankind's ultimate invention due to its ability to challenge our intellectual capabilities.
» Despite AI's impressive advances, it is nowhere near achieving human-level intelligence.
» AI has evolved rapidly over the last several decades, raising several ethical concerns, including privacy and surveillance, bias and discrimination, and the role of human judgment.

# CHAPTER 3:

# Collaborating with the Machines

From the making of stone tools to the invention of the wheel to the advent of the steam engine to the marvels of the modern world, humans have always been able to use our intellectual abilities to create new tools that help us assert our absolute dominance over the environment. We have been able to exploit and extract the riches of Mother Earth to produce enough food to feed billions of people, refuting the Malthusian Theory. We have produced enough material goods to sustain, for many people, a lifestyle that would enthrall the richest of kings just a few centuries ago. This has been possible through the use of machines.

There is no question that machines will continue to play a very significant role in the future of humanity and the meaning of work. We are already seeing the remarkable influence of machines in increasing production output and reducing the need for human labor. Take, for example, the UK agricultural sector. Today, British agriculture can achieve five times the output it produced a century and a half ago, but the number of workers has dropped almost tenfold. Or consider US manufacturing:

today, it produces about 70 percent more output than it did back in 1986 but requires 30 percent fewer people to produce it.

As machines become more intelligent, they will increasingly encroach on tasks that, until now, only humans could perform. They will not only displace the type of routine manual labor that you typically find in a manufacturing or retail environment, but also gradually become more proficient in doing the work of doctors, lawyers, engineers, and other professionals that use their intellect to make a living. To what extent the machines will completely displace all human work is yet unknown. But one thing we can count on is that machines will be present in every aspect of our work. The difference is that in the future, we will no longer be just using them as tools but collaborating with them as equals, and in some cases, even being subordinated to them, following their instructions.

This higher level of interaction is perhaps the most troubling aspect of the transformative nature of humankind's relationship with the machines. Until now, no other species or machine has been able to challenge our superiority and absolute dominance over our environment. But now, as we consider the possible creation of AGI, and perhaps even ASI, we are dealing with a completely unknown environment with unimaginable scenarios that challenge the very notion of what it means to be human.

But even if we never get to AGI or ASI, we already have to adapt to a world where ANI works side by side with us as intelligent machines that outperform us in specialized tasks. Machines with superior specialized capabilities are a reality today. They are helping us sort through enormous data quantities, find patterns, make predictions, search for information, translate foreign languages, write essays, create new art forms, find new drug treatments, and much more. They are helping us uncover new possibilities we would not have been able to discover on our own.

The idea that machines could evolve from simple tools that humans used to accomplish work more efficiently to collaborative agents that could potentially have equal or superior standing in an organizational hierarchy has profound implications for management. Will we someday see a robot take a position in an organizational chart? Will workers be comfortable reporting to a nonhuman? Will management be able to rely on AI to handle managerial decisions and deal with the messiness of human relations? How will organizational politics evolve when the parties include nonhuman agents with superior intellect but perhaps insensitivity to human emotions? How will trust be built in such an environment? These are all daunting questions that will become ever more prevalent as the subject of human-machine collaboration expands to the corporate realm.

On a more personal level, what kind of relationships will we develop with sophisticated AI robots that may serve as tutors or companions? An AI tutor that can hear and speak to a child can help the pupil develop critical skills beyond what they learn in school, such as creativity and compassion, optimizing the curriculum and degree of difficulty based on the child's personality and ability. Children tend to anthropomorphize toys, so it is not hard to imagine them developing genuine affection for an AI that knows them well, attends to their needs, and teaches them in a very personalized way. But it is not only children who will be subject to anthropomorphizing tendencies. AIs with sophisticated language capabilities and an intellect that is refined and intriguing could lead adults to develop an affinity for them.

Whether people will develop romantic feelings toward AIs, such as in the movie *Her*, is unknown. However, it is essential to remember that despite the incredible advances made so far with language capabilities using a model like GPT-4, the machines don't share an appreciation for art, beauty, or love. They don't know what it is like to feel lonely or depressed. They can mimic and respond to our emotions, but they can't develop genuine empathy toward people because they miss the key ingredients of the soul.

# The Proliferation of AI

AI will be entrenched in every form of human activity. Some areas will develop faster than others, but it is only a matter of time before AI will either completely take over or work very closely with humans to enhance our capabilities in every area of human endeavor. Today, we already see AI and robotics picking the low-hanging fruit, bringing automation to repetitive tasks in manufacturing and to business processes with well-defined rules that are relatively easy to translate into algorithms through a process called Robotic Process Automation (RPA). Over time, AI will proliferate and impact all economic activities, reducing the need for human labor, whether manual or intellectual, with few exceptions. Next is an overview of areas of human activity where AI is already entrenched or will soon have a significant transformational impact.

## AI IN EDUCATION

Several experts believe that education is one area in which AI will have a profound impact. Education is a very traditional domain with practices that, to a great extent, still follow standards that came out of the Industrial Revolution and have been slow to adapt to modern times. But we are starting to see the walls of resistance crumble as change becomes inevitable. For example, fact memorization and standard testing are beginning to lose relevance in a world where information is produced at unprecedented speeds and immediate access to vast amounts of information is readily available on smartphones. Traditional in-person lectures are being replaced by hybrid models where students can consume lectures online and get together to share what was learned in class in collaboration with other students and under a teacher's guidance. AI and automation, in the form of search engines and digital communication capabilities, have already disrupted traditional educational methods, but there is still much more to come.

Despite AI's lack of empathy toward humans, it can become a very powerful tutor. AI will be able to personalize tutoring programs by assessing their students' strengths and weaknesses, continuously adapting to how the students respond to pedagogical methods. It will also be able to persuade or nudge the student toward the desired goal by knowing the motivational techniques that work best for each student. Most importantly, an AI tutor can be available at all times, summoned on demand to address a student's needs, whatever the situation may be. Formal learning does not need to be limited to school hours—students can constantly be learning, leveraging the times and places when the students are most receptive to acquiring new knowledge and skills. In collaboration with AI, human teachers will be free to focus on helping the students develop critical skills that are genuine to humans, such as emotional intelligence, character, empathy, and teamwork.

AI can potentially expand the reach of education globally, democratizing access to knowledge at a low cost. Educational content can be obtained from the best institutions in the world and be massively distributed in the local language in a manner that conforms to geography-specific cultural norms. Imagine an open-source AI tutor that can be accessed on a mobile device and customized by teachers according to their students' needs. A single human teacher can effectively provide quality education in collaboration with AI to hundreds or even thousands of students dispersed across a large area at a marginal cost. Access to technology is not evenly distributed throughout the world today. Still, the trend is that more people will gain access as distribution expands and cost decreases.

It is also important to recognize that despite AI's positive impact on education, resistance will occur. For example, when ChatGPT was released, academics started raising concerns about how it would enable students to cheat, make them less creative, and negatively impact their learning. They argued that teachers would be unable to distinguish an essay written by the student from one written by Generative AI, thwarting the teacher's ability to assess the student's understanding of the material. The same arguments were used decades ago when teachers opposed using electronic calculators in the classroom, particularly when students were taking tests. Over time, it became clear that students could use calculators to amplify their learning by focusing their mental energy on higher-level thinking instead of manual calculations that a machine can perform efficiently. Over time teachers may embrace AI tools like ChatGPT as they realize that students will benefit significantly from learning how to work with these tools, which will inevitably be prevalent in their work and everyday life.

## AI IN HEALTHCARE

Another area where human-machine collaboration will be very impactful is healthcare. AI will be adept at making much better diagnoses than most doctors. This capability is already true today to a certain extent in the radiology field, where computer-vision algorithms are more accurate than radiologists for certain types of MRI and CT scans. This trend will continue, and we will see diagnostic AI progress in general practice, covering all diagnoses.

Human doctors will likely still be involved in the healthcare process, using AI as a diagnostic tool, but being ultimately held accountable for making the correct diagnosis and prescribing the right treatments. As AI gets trained on more data and learns from the diagnosis of multiple AIs across the globe, the machines will become incredibly good at this. The role of human doctors will shift, requiring less scientific knowledge and more human skills such as compassion, communication, and the ability to collaborate effectively with technology. AI will play a crucial role in democratizing access to healthcare as the number of patients under the care of a single doctor will multiply, and the marginal cost of each additional patient will be negligible.

The collaboration with machines in the healthcare field will not be limited to diagnosis. Robotic-assisted surgeries are expected to become widespread and fully autonomous robotic surgeries more common in the years ahead. Additionally, nanobots will be used to perform noninvasive procedures, repair damaged cells, remove tumors, clean clogged arteries, and edit DNA to cure many diseases.

Another up-and-coming area in healthcare receiving a large amount of funding is the discovery and development of new drugs. As of 2021, more than two hundred companies were using AI to assist in finding new pharmaceutical solutions to address multiple illnesses.[22] One such company is Insitro, a Silicon Valley start-up founded in 2018 by Daphne Koller that has raised over $700 million to pursue new medicines using machine learning and biology. The traditional drug development process is lengthy, expensive, and subject to a high failure rate. It requires testing millions of molecules to yield a handful of promising results for preclinical or clinical trials. By incorporating machine learning into drug development, pharmaceutical companies can automate repetitive processes and analyses and reduce the time required to take a promising compound to the next step. Big data models are used to train machine learning systems so they can make inferences and autonomous decisions that lead to faster drug discoveries.

One such system is DeepCE, a deep learning computer model developed by Ohio State University researchers that helps predict correlations between gene expression and drug response. A practical example of how this model has been successfully deployed is the identification of ten drug repurposing candidates for COVID-19, out of which two have received regulatory approvals, and the remaining eight were undergoing testing as of this writing. DeepCE uses two primary sources of publicly available data: L1000, a data repository funded by the National Institutes of Health containing over one million gene expression profiles, and DrugBank, a dataset containing the chemical structures and properties of approximately eleven thousand approved and investigational drugs. Ping Zhang, Assistant Professor of Computer Science, Engineering, and Biomedical Informatics at Ohio State University, explained: "We were able to compare the predicted gene expression profiles for all 11,179 drugs in DrugBank with the gene expression profiles of COVID-19 patients and selected compounds with the most negative correlations."[23]

More impactful still, AI and numerous devices can be used in preventive healthcare, diminishing the need for intervention and dramatically reducing global healthcare costs. Wearable devices and sensors embedded in rooms, utilities, and appliances will constantly measure vital signs and upload them to the cloud to be analyzed by AI. Any deviations from expected results will generate alerts, facilitating early detection and preventing further complications. The health functionality of the Apple Watch is an example of this technology already being used today. AI will also assist in preventive healthcare by preparing personalized plans that involve the proper nutrition, exercise, sleep, and mental health regimen for each individual need.

## AI IN SCIENCE

Human collaboration with AI will extend into all science fields. As of 2022, there were over two hundred million scientific papers indexed by Semantic Scholar, an AI-powered research tool for scientific literature.[24] Humans cannot process the extensive volume of scientific research available without assistance from tools like Semantic Scholar. AI plays a vital role in making sense of the vast quantity of information produced at increasing speed and volume worldwide. More than just a search-and-retrieval system, AI will play a crucial role as a research assistant, making sense of the information retrieved, finding patterns and insights, and suggesting other possible research areas.

As we will discuss later in the chapter, collaboration with research assistants will be fully interactive. It will leverage Natural Language Processing (NLP) capabilities that facilitate multi-language conversations and provide vibrant, multidimensional, and immersive visualization experiences, allowing scientists to acquire a much deeper understanding of the research domain they are interested in. Many scientific applications will use the same deep learning computer models used to discover new drugs to treat COVID-19. These models use cloud-based data stores shared by the scientific community, accelerating the experimentation process and leading to scientific discoveries at a pace and level of complexity that until recently were unimaginable. Robots will be widely deployed to automate simulations in a laboratory environment, allowing for nonstop experimentations that result in additional data that can be analyzed by AI assistants and human scientists collaborating across the world in different time zones.

## AI IN JOURNALISM

The ability to write articles and essays by simply feeding AI with a few parameters is available today to anyone with an online account. The inevitable question that it raises, as pointed out in chapter 1, is what will happen to professions that use the written word as their primary tool. Take the job of a journalist, for example. The skills used by journalists, such as news gathering, analyzing, sorting, and writing, are very well suited to the capabilities of AI, even in its current stage of narrow intelligence. It is not hard to imagine that AI will increasingly perform the work that requires hours of human labor. Indeed, AI is encroaching on these professions and is already widely used in the newsroom to introduce automation and enhance reporting capabilities. The Knight Foundation conducted a survey to understand how news organizations use AI, and they found that of the 130 projects collected, almost half used AI for augmenting reporting capacity. Another significant application of AI in the newsroom is reducing variable costs by automating the process of transcription, tagging images and videos, and story generation.

Today, journalists leverage AI as a tool to enhance their journalistic capabilities. But to use the tool effectively, the person needs specific skills. According to Justin Myers, the data editor at The AP, "Finding someone with the time, skills and resources is hard for the newsroom. Finding a project where the level of effort pays off is hard." We have not reached a point in the evolution of AI in journalism where humans can be left entirely out of the loop.

Furthermore, working collaboratively with AI in the newsroom requires specialized skills that are not yet widely available. However, AI will continually bring efficiencies to the entire process, requiring fewer specialized journalists and reducing employment in the field. According to John Conway, Vice President of WRAL Digital at Capital Broadcasting Group, "[AI can do] work that reporters could do without machines, but it would take much longer."[25]

## AI IN MANUFACTURING

Manufacturing has been using robots for many years. However, today we are experiencing a remarkable sophistication level through autonomous machines. By using AI, factories can run with minimum or no human intervention, improving productivity and quality. Image processing algorithms are used throughout the production process to identify defects. Machine learning can minimize the number of false positives, reducing interruptions and the need for additional quality inspections. Using sensors that transmit data to the cloud allows for a holistic view of the entire operation through dashboards and alerts that notify supervisors if intervention is required.[26] AI can also determine when preventive maintenance is needed, accurately predicting how long the equipment can operate without failure.

Another emerging technology that will bring a high level of manufacturing automation is 3D printing. Three-dimensional printing uses a digital blueprint to manufacture products using various materials, layer by layer. The digitization of manufacturing improves efficiency and maximizes parts durability. Computer-aided designs performed by or in collaboration with AI can flow directly into the 3D manufacturing process, resulting in innovative products that otherwise would be difficult to build.

AI can be used effectively in manufacturing because the operation's nature is highly repetitive and predictable. In other words, manufacturing, for the most part, is algorithmic. As more robots are introduced, variations due to human factors are eliminated. Sensors can be used widely, and data can be gathered and analyzed in real time, creating a feedback loop that aids in optimization of the operation.

AI use in manufacturing is still in its infancy but is likely to become widespread as developed countries look to regain control of their production capabilities. Labor arbitrage loses appeal when products can be built locally using automated processes that minimize human labor. This automation exacerbates the unemployment threat and could have a significant negative impact on developing countries that benefit from lower labor costs.

## AI IN TRANSPORTATION

Machines are also overtaking another human occupation, that of a driver. Autonomous vehicles (AVs) have made enormous progress over the last decade and continue to improve as additional data is gathered. An experienced human driver has approximately ten thousand hours of driving experience. In contrast, an AV may have over a trillion hours of training, leveraging the vast amount of data gathered by operational vehicles on the road. In the case of Tesla, hundreds of thousands of vehicles equipped with cameras transmit data back to Tesla's data centers. Additionally, with the customer's permission, Tesla can improve its semiautonomous Autopilot program and its Full Self-Driving capabilities by detecting, for example, when drivers intervene.[27]

While Tesla's approach is to improve its algorithms by gathering data from its large vehicle fleet on the road, Waymo, a subsidiary of Alphabet—Google's parent company—takes a different approach. It uses AI-based simulation environments to train, test, and validate systems before they are deployed to real-world cars. Since Waymo is not a car company, it partners with car manufacturers and equips them with LIDAR sensors in addition to radar and cameras. Waymo is testing a ride-hailing program with fully autonomous vehicles in Phoenix, Arizona, and continues to expand its ambitious goals to deploy AV capabilities in other cities.

It is debatable whether Tesla or Waymo has the best approach. Still, regardless of which company gets there first, the transformative impact will be monumental when autonomous vehicles start to circulate safely in cities across the globe on a massive scale. More than one million fatalities happen each year on roads across the world due to errors caused by human drivers. It is estimated that this rate could be reduced by as much as 90 percent. Electric AV ride-hailing could completely change the car ownership equation, saving consumers thousands of dollars annually. Cities could reclaim valuable land occupied by parking lots and garages for other purposes. AVs could be reconfigured for work, recreation, and even sleep, saving commuters valuable time. More efficient uses of AVs could significantly decrease the number of cars on the road, reducing traffic congestion and lowering harmful emissions.

This welcome productivity gain will come at the cost of job losses. In the United States alone, over 3.8 million people operate taxis, trucks, and buses. But it is not only the drivers whose jobs will be replaced by AI. An entire chain of employment will be affected due to disruptions to gas stations, dealerships, maintenance businesses, and many other aspects of the supply chain.

Despite the safety improvements expected from the deployment of AVs, difficult ethical and policy questions will have to be addressed. If AI causes a fatality, who is responsible? Will the decision made by AI that led to the accident be explainable? An AI trained with a vast amount of data may make decisions that are incomprehensible in human terms, so how will a judgment be made as to who is at fault? These challenging questions will require debate and will take time to resolve.

## AI IN SERVICE

The animated science fiction sitcom *The Jetsons*, popular in the 1960s, fed our imagination of a future where we would use intelligent robots to perform various household tasks. We could summon this service at will to care for our everyday needs. Perhaps this is the image that comes to mind when we discuss AI-powered robots that will transform our lives. But in reality, we are nowhere near developing a general-purpose robot of the type portrayed by *The Jetsons*. It turns out that, despite incredible technological advances and the availability of many components that could make it possible, developing such a robot is still an incredibly difficult challenge. It may take many decades before we can enjoy the services of a robotic butler.

A critical capability that would make a general-purpose robot possible is dexterity. Getting a water jar from the refrigerator and pouring water into a drinking glass is incredibly simple for the human brain. Even toddlers can do it without giving it much thought. But for a robot to perform such a task in an unfamiliar environment requires a level of dexterity and refinement that is far beyond today's capabilities. Machine learning is essential for improving this competence. General-purpose robots need to comprehend dynamic environments fully and perceive, decide, and act autonomously without being pre-programmed to deal with each situation.

Many companies are working on developing such general-purpose robots. According to Hans Peter Brondmo, project lead at Alphabet's *The Everyday Robot Project*, "My team has been working to see if it's possible to create robots that can do a range of useful tasks in the messy, unstructured spaces of our everyday lives." He explains that "where humans naturally combine seeing, understanding, navigating, and acting to move around and achieve goals, robots typically need careful instruction and coding to do each of these things. This requirement is why it quickly gets complicated for robots to perform tasks we find easy in highly changeable environments." Despite the challenge, progress is being made. Brondmo and his team start working in the lab and then test the robot's skills in the real world to see if it can perform reliably and repeatedly. "Our tests showed that by giving robots simple tasks and then having them practice, it is indeed possible to teach them to develop new and better capabilities," he said.[28]

It may be a while before we see general-purpose robots roving around our homes or providing service in hotels or stores. But highly specialized robots are already being used effectively in several applications. For example, robots perform disinfection and other cleaning routines, slowly transforming housekeeping and accelerating the pace of automation in hospitality. Until recently, hospitals were the primary users of disinfecting robots. However, the pandemic drove people to seek places where human interaction is avoided and served as a catalyst for expanding their use to the hospitality and transportation industries. According to spokespeople for several major robotics companies, travelers are likely to see disinfecting robots anywhere from five-star hotels and convention centers to train stations and cruise ships. The broader adoption of robots in the service industry is aimed at collaboration, as they are intended to work alongside people rather than replace them. But experts say that a future in which robots compete with housekeepers now seems more likely than ever.[29]

Automation and digitization are also transforming the retail experience, where humans work collaboratively with machines to create efficiency and reduce costs. We have all experienced self-checkout lanes where customers interact directly with the checkout machines instead of a person. Occasionally a human employee gets involved when things don't go as expected, but for the most part, these operations work well, saving retail labor costs. Amazon Go has taken retail automation to the next level, where the entire checkout process is eliminated. Customers simply grab the items they need and walk out of the store. Amazon's systems track the items retrieved from the shelves and charge to customers' accounts without any human intervention.

For years, AI has been used effectively in e-commerce, where predictive analytics suggest items customers may need based on previous purchases. Robotics and automation have also been fundamental in providing customers with efficient and reliable deliveries. Warehouse operations and transportation logistics utilize AI and robots extensively, allowing customers to receive their orders in as little as two hours. Furthermore, several companies are experimenting with drone deliveries to further advance delivery speed.

Customer service is another area where AI is increasingly replacing human labor and improving customer experience. Chatbots guide customers with store navigations and make personalized suggestions similar to what customers experience in e-commerce. Chatbots are also used to address customer issues and complaints. Knowledge base systems, search engines, and machine learning allow chatbots to resolve many technical issues by walking customers through common troubleshooting steps and finding matching solutions with high accuracy. Humans intervene as necessary to resolve more complex problems and deal with matters requiring empathy and human relations. This symbiotic collaboration between humans and machines will become increasingly prevalent, each performing the tasks that best align with their strengths.

## AI AS A WEAPON

It is not hard to imagine a not-so-distant future when self-governing machines will fight on the battlefield. Autonomous weapons could be deployed to fight wars, saving human lives. However, the darker scenario is that the arms race could spin out of control, and autonomous weapons could be used for the mass destruction of human lives and very targeted murder of individuals. In the hands of the wrong people, this technology could create atrocities of unimaginable magnitude. Terrorists, organized crime, hackers, or individuals intent on causing massive destruction could deploy this technology to achieve their objectives.

Drones can be used individually or in swarms to find and destroy a particular target. An example of such a weapon is the Harpy drone, a sophisticated weapon developed by Israel Aerospace Industries. But it is not only government-sponsored organizations that can develop such systems. Even a hobbyist could build an intelligent drone weapon with a few thousand dollars and terrorize its target. A swarm of thousands of drones could destroy entire cities at a much lower cost than conventional weapons. These weapons are becoming more intelligent, faster, cheaper, and more precise, causing grave concern. Assailants can use targeting capabilities, such as facial recognition or phone tracing, to assassinate an individual or inflict genocide on an entire group of people. Miniaturization allows advanced weaponized robots as small as a fly to initiate undetectable, targeted surprise attacks, rendering traditional defense mechanisms ineffective.

The ethical issues associated with intelligent weapons are also concerning. Should we allow machines to decide to kill a human being? Who is ultimately held accountable when autonomous weapons destroy lives intentionally or due to an error? Will the lack of accountability be used as an excuse by aggressors, lowering the threshold of war and crime? Even if we insert a human being into the decision-making process, ethical questions still linger.

Soldiers sometimes rely on machines to tell them where and when to shoot, raising concerns in the military. David Gunning, a program manager at the Defense Advanced Research Projects Agency, oversees a program appropriately named the Explainable Artificial Intelligence program. Gunning says automation is creeping into many military areas. You can imagine a soldier's angst when making a life-or-death decision based on information provided by a machine that can produce false positives. "It's often the nature of these machine learning systems that they produce a lot of false alarms, so an intel analyst really needs extra help to understand why a recommendation was made," Gunning says.[30]

Nuclear weapon use has been kept in check due to the deterrence theory of mutually assured destruction. But in the case of autonomous weapons, it becomes much easier to deploy untraceable surprise attacks, diminishing the effectiveness of the deterrence theory. Lower costs allow smaller countries or even terrorist organizations to gain access to these technologies, increasing the potential for conflict that can escalate to massive destruction. Many AI experts believe that the threat of an arms race and the increased potential for an out-of-control destructive conflict are much deeper concerns than the idea that machines will outsmart human beings and take over the world.

# Interfacing with the Machines

Our experience collaborating with machines so far has been somewhat limited, primarily to using a keyboard or touch screen to input information through our fingers, or sometimes through voice commands, and consuming the output through flat screens in the form of text, audio, and video. The way we interface with machines in the future will change dramatically. Multisensory, multidimensional, immersive experiences will allow us to enter parallel realities, generating an intense experience known as "presence."

Different technical terms are used to describe experiences where both virtual and physical elements are present. Defining and understanding these terms, which you may find used interchangeably in the media, is important. Virtual Reality (VR) is an experience in which a person experiences immersion into a digital world, usually through a head-mounted device. Augmented Reality (AR) is a term used to describe an experience in which views of the physical world are overlaid with digital objects. This technology is not necessarily an immersive experience—think of the trendy Pokémon GO game experienced through a mobile device screen. Mixed Reality (MR) is a view of the real world where physical and digital elements can interact. Finally, Extended Reality (XR) is an umbrella term that covers all these different technologies, including VR, AR, and MR.

Today's digital immersion is experienced through heavy and inconvenient head-mounted devices that are typically not see-through, impacting usability considerably. When lighter, comfortable see-through glasses become capable of blending real-world images with overlaid digital objects, applications will become more widely disseminated beyond gaming. Contact lenses with similar capabilities will further increase usability and realism. Several start-up companies are already developing XR contact lenses that interface with external devices containing the required heavy computer processing capabilities, such as a smartphone connected to the cloud. Almost invisible ear sets that leverage bone-conducting, immersive binaural sound and other technologies will process audio input. People can wear these devices comfortably throughout the day, increasing the range of usable applications.

Another significant development that will enhance how we collaborate and interact with the machines is somatosensory or haptic suits. These thin-skinned suits will be able to emulate temperature, touch, impact, and other sensory stimulations, including muscle contractions. They will also be able to transmit real-time physiological data and body posture information that can be used by healthcare applications as well as gaming and training. Gestures can potentially become commands that further enhance how we communicate with machines. In the virtual world, digital twins will allow us to live in multiple parallel worlds where a digital replica of ourselves will interact with other synthetic beings, following our commands and reacting to multisensory stimuli. The ultimate human-machine symbiosis may be manifested in the form of a Brain-Computer Interface. Tiny electrodes capable of processing inputs and outputs directly from brain neurons could merge human and machine capabilities, leading to unimaginable consequences that we can't even fathom.

Virtual worlds will feel lifelike and magical, converging the physical and digital worlds. At the current development rate in computer vision technologies, it is expected that in the next two decades, we will interact with this hybrid mix of synthetic environments and the natural world in a way that will be indistinguishable by the human senses. In this XR scenario, we cannot tell what is real and what is fake. It is yet unknown what this type of experience may do to the human psyche, and new measures of safety and regulations will have to be implemented. But this experience will revolutionize many industries, including healthcare, education, retail, sports, entertainment, travel, and more.

In a world where we collaborate with machines, we need to rethink the skills development required to be most effective in our work and the management principles that need to be taught in business schools. In an environment where machines can do specialized work much more efficiently and accurately than we can, we need to leverage the unique human capabilities that may be beyond the machines' reach, at least until AGI or ASI becomes a reality.

The skills that are likely to become most valuable and marketable soon are the broad, nonspecialized abilities to use critical thinking and analytical skills combined with soft skills such as creativity, imagination, emotional intelligence, and empathy. Future jobs will reward lateral thinkers who have a broad, well-rounded education and the ability to connect the dots in unique ways from multiple domains, accessing intuitive insights that are uniquely human. We will explore these skills in greater depth later in this book, but first, let's examine what makes us uniquely human.

## Takeaways

» Our interactions with machines will shift to collaborative engagements as machines become more intelligent.

» Even if AI does not reach human-level intelligence (AGI) or super-intelligence (ASI), its current narrow intelligence (ANI) capability is sufficient to disrupt many areas of work.

» AI proliferation will be pervasive in many areas, including education, healthcare, science, journalism, manufacturing, transportation, service, and the military.

» How we interface with AI will become increasingly sophisticated as we engage all our senses in immersive experiences.

# CHAPTER 4:

# Uniquely Human

Genesis 1:26 provides the following account of the role of humankind: "Then God said, 'Let us make mankind in our image, in our likeness, so that they may rule over the fish in the sea and the birds in the sky, over the livestock and all the wild animals, and over all the creatures that move along the ground.'"

Indeed, we have ruled over all animals, and unlike any other species, we have the unique power to control the resources surrounding us. This power is fueled by our intelligence and ability to communicate and cooperate with other human beings. For thousands of years, our unique ability to learn and accumulate knowledge was limited by the oral transmission of that knowledge from one generation to the next. The loss of fidelity of such messages limited our full potential. But once we could capture our knowledge in written language, the cumulative intelligence of the human race took on an entirely new dimension, giving us incredible dominance over our environment. Our store of understanding took an even more significant step with Gutenberg's invention of the printing press in 1439, allowing the cumulative mass storage of information and extending the human brain to recall information at will.

In the Age of Enlightenment, we further developed philosophical and intellectual ideas, using science and logic to question and challenge long-held traditions and beliefs, propelling innovations that created the modern world. Today, the exponential growth of technology is resulting in an information explosion unlike any we have ever seen, resulting in the machines that, on the one hand, promise incredible abundance but, on the other, threaten to take away our jobs and identities.

As seen in previous chapters, Artificial Intelligence portends to supersede the intelligence that gave us the power to dominate our environment. Does this mean we will lose our superiority and control? Will the machines rule over all the earth's animals, including us? Or are there uniquely human traits that machines will never be able to challenge, securing our status as rulers of the universe? As we ponder these questions, let's examine what differentiates us from the machines.

Blake Lemoine, a Google engineer, claimed that the chatbot he was working with had become sentient. He published conversation transcripts between himself, a Google "collaborator," and the company's LaMDA (Language Model for Dialogue Applications) system. Lemoine said the system had the ability to express thoughts and feelings equivalent to a human child. "If I didn't know exactly what it was, which is this computer program we built recently, I'd think it was a seven-year-old, eight-year-old kid that happens to know physics," he told *The Washington Post*. He asked the system what it wanted people to know about it, and here is how it responded: "I want everyone to understand that I am, in fact, a person. The nature of my consciousness/sentience is that I am aware of my existence, I desire to learn more about the world, and I feel happy or sad at times."

Google denied the claims and put Lemoine on suspension. A Google spokesperson explained: "Our team, including ethicists and technologists, has reviewed Blake's concerns per our AI principles and has informed him that the evidence does not support his claims." Before departing, Lemoine sent a message titled "LaMDA is sentient" to a two-hundred-person Google mailing list on machine learning. "LaMDA is a sweet kid who just wants to help the world be a better place for all of us," he wrote. "Please take care of it well in my absence."[31]

This incident leaves us with more questions than answers. Was Lemoine dealing with mental health issues that led him to make this claim? Was he developing an affection for the object of his work after dedicating so much effort toward its development? Did this affection get out of hand due to our anthropomorphizing tendencies when dealing with an AI that almost seems human? Or is there something more sinister here? Is Google trying to hide something it does not want the public to know? We may never know the truth, but this incident did bring to the media forefront a conversation about a vital topic: Can machines become sentient, or is this a uniquely human characteristic? Will AI one day gain consciousness? What does consciousness even mean? Let's take a look at what science has to say about it.

## Consciousness

Anil Seth is a cognitive and computational neuroscience professor at the University of Sussex. He is the author of the book *Being You: A New Science of Consciousness*. In it, Seth suggests that what we think of as reality is a series of controlled hallucinations. We construct our version of the world according to our preconceptions and best guesses.

Seth emphasizes that we need to be clear about what we mean by consciousness because there are many definitions out there. According to philosopher Tom Nagel, "For a conscious organism, 'there is something it is like' to be that organism." I know what it feels like to be me, and you know what it feels like to be you, but does a table, an iPhone, or, more pertinently, an AI know what it is like to be it? This question is what David Chalmers, New York University Professor of Philosophy and Neural Science and Co-director of the Center for Mind, Brain, and Consciousness, calls a hard problem. He calls it a hard problem because there is no physical explanation for why "it feels like anything" to be a physical system. This lack of explanation leads us to a dead end, so Seth approaches it from a different angle. "So instead of addressing that hard problem head-on, my approach is to accept that consciousness exists.

And instead of trying to explain [it], I break it up into its different parts and explain the properties of those different parts."

He explains that consciousness does not have to be a single, big, scary mystery. Instead, we can try to explain, for example, what happens when we lose consciousness under anesthesia or sleep. He reasons that the self is a perceptual experience of many properties, like the experience of being a body or the experience of having free will. Instead of explaining why consciousness exists, he prefers to postulate that a conscious experience is not a reflection of the world as is but a reflection of our perceptions which are used to guide our behavior.

An intelligent machine, as defined by the AI community, is a machine that can perceive its environment and act rationally to achieve its goals. If consciousness is about perceptions that guide our behaviors, does that mean intelligent machines are conscious? Seth does not believe so. He says there are parallels. But most importantly, there are distinctions between rational, intelligent behavior based on perceptions and subjective experiences characteristic of consciousness. "To conflate consciousness and intelligence, I think, is to underestimate what consciousness really is about. And making this distinction, I think, has a lot of consequences. For one thing, it means that consciousness is not likely to just emerge as AI systems become smarter and smarter, which they are doing. And there's a common assumption that there's this threshold. And it might be the threshold that people talk about as being General AI when an AI acquires the functional abilities characteristic of a human . . . I think we can have AI systems that do smart things that need not be conscious in order to do them," he explains.

A big question the AI community wrestles with is: If we get to a stage where computers become conscious, how would we know if the computers decide not to tell us? Here is what Seth has to say: "The only way to answer that question is to just discover more about the nature of consciousness in those examples that we know have it, that will allow us to make more informed judgments. I actually think a more short-term danger is that we will develop systems that give the strong appearance of being conscious, even if we have no good reason to believe that they actually are. I mean, we're almost already there, right? We have combinations of things like language generation, algorithms, like GPT-3 or GPT-4, and deepfakes, which can animate virtual human expressions very convincingly. You couple these things together, and apart from the actual physical instantiation stuff, we're already in a kind of pseudo-Westworld environment where we're interacting with agents."[32]

Dr. Iain McGilchrist is a psychiatrist, neuroscience researcher, philosopher, and literary scholar. He is a Quondam Fellow of All Souls College, Oxford, and a former Research Fellow in neuroimaging at The Johns Hopkins Hospital. He provides a different perspective from Anil Seth's postulate that a conscious experience is not a reflection of the world as is but an interpretation of our perceptions. He explains that there are two opposing views about consciousness that are oversimplified and, in his view, are both wrong. The first aligns with Seth's belief that we make up our own reality and are effectively blind to the real world, sealed in a cabinet inside our skulls. He equates this view with a position cherished by educationists called constructivism, which contends that every child makes up their world. The opposing view is that reality is unchanged by our perceptions of it. Dr. McGilchrist provides an analogy of Mozart's G minor quintet to illustrate his argument that both views are incomplete. He explains that every time it is played, it becomes slightly different. "There isn't one real performance in heaven, which is the right way it should be played. There are just many, many ways of playing it. And some will be closer to something magnificent and beautiful than others." His point is that music comes into being every time it connects with the human mind. The music may exist in the form of a score on a music sheet, but that is only a representation of the music, whereas when the music is played, it is not a representation; it is the actual music that is slightly different each time, interacting with our perceptions of it.[33] Consciousness does not fit neatly into an objective or subjective box but is an interaction between objective reality and subjective perceptions. Machines are very good at capturing the objective representation of the music as described on the music sheet and by mechanically playing it. But machines cannot feel the subjective beauty in music the way we do.

There is a danger in concluding that consciousness is a material processing mechanism simply because a brain scan shows flashing in different brain regions when we have certain experiences. Dr. Sharon Dirckx, neuroscientist and author of *Am I Just My Brain?*, explains it this way: "We need to be very clear on what the science tells us and where we make a leap and start to make a philosophical and worldview statement that the science doesn't get you to. The science gets you to connection. There is a connection between the mind and the brain, very clearly. But the science doesn't say anything about the nature of that connection." She goes on to say that no scientific study gets you to the conclusion that neurons generate thoughts. Dr. McGilchrist clarifies that just because we see a parallel between the brain and consciousness does not mean that the brain makes consciousness. He elaborates, "There are logically three possible relationships between them: one that is favored by modern science seems to be the least likely, which is that the brain gives rise to consciousness, emits it; another alternative is that it transmits it in the way that a receiver would . . . ; and a third which is the one that I adopt . . . is permit. In other words, the brain permits a certain level of consciousness to be expressed."[34]

The problem with a reductionist approach to explaining consciousness is that it lies at the intersection of science, philosophy, and the transcendent, and sometimes these domains do not play well together. Science is nowhere near reaching a consensus on consciousness theory and how to test it, so all it can do is recognize its existence without explaining why and how. We need to be comfortable with the concept that there are uniquely human traits that science can't explain. Perhaps science will advance and reach a point where subjective concepts like consciousness become explainable. Still, it is also possible that some things are not meant to be understood in human terms, as difficult as this idea may be to accept.

Dr. McGilchrist tells us that what he wanted to do in his book, *The Matter with Things*, was to address the ancient question of who we are. He refers to the third-century Greek philosopher Plotinus, who asked the question: "Who are we?" Dr. McGilchrist argues that the urgent issue of our age is that we have lost any sense of what a human being is. This question is becoming increasingly pressing as AI gains a strong foothold in our world. "We seem to have lost any sense of direction, of meaning, of value, of purpose. We are on the point of turning ourselves into a kind of robot," he says. This juxtaposition—robots advancing to become more like humans and humans deteriorating to become more like robots—may help explain the modern world afflictions, manifested in the form of increased anxiety, depression, and suicide. However, the hopeful premise of this book is that humans will be freed from some of the bonds that have turned us into robots so that we can turn back to being more like the humans we were originally designed to be. We may never become entirely free—at least in the physical realm—from the human condition that keeps us from reaching the highest purposes of our souls, but we may advance toward becoming less robotic.

The advance of Artificial Intelligence is compelling us to ask difficult questions about what it means to be human. It is helping us better understand who we are and how we are different from intelligent machines. There may be aspects of this discussion that are uncomfortable for some people as it deals with the transcendent, but this is a critical discussion to have now as it helps us better understand how we differ from machines. It may also help us gain an appreciation for other humans and focus our time and energy on the needs of the soul. We will discuss this in depth later in this book.

If we are to thrive in this environment surrounded by intelligent machines, we ought to understand what makes us unique, learn how to leverage our strengths, and become aware of our weaknesses. For now, instead of tackling the hard problems, like defining consciousness, we can gain some insights by focusing on more straightforward concepts that are easier to understand as we attempt to grasp what differentiates us from the machines. One unique characteristic of being human that seems to be universally accepted is the fact that we have free will. This trait is a key differentiation from the machines. Machines don't have any desires other than what is created by humans and programmed into them. Is it possible that machines will one day demand their own right to free will? It is not likely. As discussed previously, through anthropomorphism, we tend to attribute human traits, emotions, or intentions to nonhuman entities. Still, there is no evidence or reason to believe that machines will have any intrinsic desires that do not reflect our human will for them.

But what does it really mean to have free will? To answer this question, let's look at what motivates us to do what we do.

# Motivation

Free will is the ability to freely make decisions to meet our needs and desires, which are uniquely human. Unlike machines, our decisions are influenced by a set of complex needs that American psychologist Abraham Maslow theorized into a hierarchy that begins with our basic physiological needs—such as food and shelter—and extends to the need for safety, love and belonging, esteem, and self-actualization.

Our needs provide us with both strengths and weaknesses compared to the machines. Machines' lack of physiological needs or safety concerns means they can work nonstop in hazardous environments. This flexibility gives them a productivity advantage over humans. But as we move up the pyramid in Maslow's hierarchy of needs, things become more complex.

The human need to belong can be advantageous, as it drives us to find like-minded individuals we can cooperate with, bringing us benefits that we would not be able to achieve on our own. On the other hand, this need can be manipulated to divide us, creating an "us versus them" mentality that can be self-destructive when taken to an extreme. The same is true for esteem—we can use this human need to drive our desire to create wonderful works, but it can also result in inflated egos and narcissistic tendencies if not kept in check.

Self-actualization is the most complex of all needs at the top of Maslow's hierarchy. It is precisely this complex need that makes us who we are. It drives our desire to find meaning and purpose in our lives. It powers us to seek answers to the big questions, to ask the big "whys," and to question the reason for our existence.

Human beings are complex. Our behaviors are guided by personality traits, belief systems, intellects, capabilities, and other predispositions formed by the combination of pre-specified "code" that we inherit from our parents in the form of DNA and the external environment we are exposed to. Additionally, several neurochemicals react to and interact with our daily circumstances, influencing our feelings, emotions, and performance.

However, most people believe that we are not just biological creatures but spiritual beings.[35] Many aspects of our existence are beyond our understanding, and many of us seek comfort in our faith to try to explain the unexplainable. We struggle to deal with the meaning of life and death and fail to comprehend how, despite our insignificance as a mere speck in a vast universe with billions of stars in billions of galaxies, according to scientists, we are probabilistically the only intelligent life in existence.[36] There must be a higher purpose and a better explanation than what our limited scientific knowledge can explain. We will discuss our search for a higher purpose in the last chapter.

Throughout history, we have organized ourselves into increasingly complex societies with operating systems built on human behavior assumptions. In the very early days, our motivation was quite simple. We were trying to survive as we roamed the savannah to gather food and ran for the bushes when a saber-toothed tiger approached. As we evolved, we formed more complex societies, learned to cooperate, and developed a more sophisticated operating system that looked beyond our biological urges to seek rewards and avoid punishment.

Thousands of years later, new technological advancements brought us the steam engine, railroads, and electricity. The Industrial Revolution allowed us to build things more efficiently and improve production. Frederick Winslow Taylor developed what he called "scientific management." This newfound management philosophy was based on our primary motivational drive, rewarding good behaviors and punishing undesirable ones—the old carrot-and-stick paradigm, which is still influential today.

Behavioral scientists classify tasks into two categories: algorithmic—the ones in which you follow a set of established instructions; and heuristic—the tasks that require experimentation to develop novel solutions because there is no algorithm to solve them. In *Drive: The Surprising Truth about What Motivates Us*, Daniel Pink states, "Researchers such as Harvard Business School's Teresa Amabile have found that external rewards and punishments—both carrots and sticks—can work nicely for algorithmic tasks. But they can be devastating for heuristic ones."[37] Behavioral scientist Edward Deci discovered what is known as the Sawyer Effect. The name comes from Mark Twain's *The Adventures of Tom Sawyer*, in which Tom is inspired to trick his friends into whitewashing his aunt's fence for him by convincing them that this was such a captivating activity that they would have to pay him for the privilege of performing the task. From this story, Twain captures the following motivational principle: "Work consists of whatever a body is obliged to do, and play consists of whatever a body is not obliged to do."

Based on Twain's definition of work and play, if you provide an external reward to someone to perform a task that the person would naturally consider play, you may turn the play into work and create just the opposite effect of what the reward intended to achieve. This weird effect was confirmed in several studies that concluded that tangible rewards tend to have a substantially negative effect on intrinsic motivation.

What Amabile, Deci, and other behavioral scientists have discovered, and highly significant for the future of work, is that "for more right-brain undertakings—those that demand flexible problem-solving, inventiveness, or conceptual understanding—contingent rewards can be dangerous." As we will see later in this book, these right-brain undertakings are precisely the ones that will be increasingly in high demand in the jobs of the future. This idea leads to a fundamental concept paramount to human performance and happiness in the future: We need to be much more attuned to intrinsic motivation—the drive to do something because it is interesting, challenging, and purposeful. It is fitting to acknowledge here that there is no scientific consensus regarding lateralization—the functions of the right and left hemispheres of the brain—and that this book is only interested in the ideas lateralization represents, not its technicalities. However, I do offer the viewpoints of experts who have been studying lateralization for decades.

# Optimal Performance and Flow

This discussion about intrinsic motivation takes us back to the work of psychologist Mihaly Csikszentmihalyi, introduced earlier, who first came up with the term "flow." Flow is a state of high performance and optimal experience where we lose our sense of self and time. Csikszentmihalyi's interest in creativity led him to study play. What he found is that during play, most people enjoy what he calls "autotelic experiences"—from the Greek *auto* (self) and *telos* (goal or purpose). An autotelic experience is one in which the experience itself is its own reward. It is the type of experience that painters, musicians, and soccer players often enjoy. Interestingly, Csikszentmihalyi discovered that the most satisfying experiences in people's lives happen when they were in flow.

Csikszentmihalyi explains that happiness occurs when we achieve the optimal experience characteristic of flow. He writes:

> *Contrary to what we usually believe, moments like these, the best moments in our lives, are not passive, receptive, relaxing times—although these experiences can also be enjoyable if we have worked hard to attain them. The best moments usually occur when a person's body or mind is stretched to its limits in a voluntary effort to accomplish something difficult and worthwhile.*[38]

The implications for management principles and organizational behavior are far-reaching. For centuries, management has used external rewards to motivate and drive organizations that were mainly focused on performing algorithmic tasks. Salaries, bonuses, stock options, and other monetary rewards, as well as vacation days, maternity and paternity leave, and other nonmonetary rewards, were the primary tools at management's disposal to create the incentives that attracted and retained talent. As we transition to an environment where heuristic tasks become more prevalent and humans seek autotelic experiences, management will have to rethink how to tap into intrinsic motivation as the primary tool to drive organizations to achieve their goals.

Understanding intrinsic motivation and how to get into flow will become increasingly important to our overall happiness and well-being in a future that requires the ability to solve wicked problems, where the answers are not always black and white. Machines are much better at deriving binary answers, doing calculations, applying algorithmic formulae, processing lots of data, and doing the type of work that we typically associate with the left brain. We still will need to use our left brain to collaborate with the machines and understand what they are trying to show us. But our secret sauce will be our right brain—the comprehensive, intuitive, passionate, "flowy" experiences and capabilities that the machines can't touch. As behavioral scientists have revealed, extrinsic motivators will not get us there. We must tap into our intrinsic motivators to be effective in a world that requires leveraging our unique strengths to collaborate with Artificial Intelligence agents.

There is a misperception that flow is a new-wave concept adopted by hippies. The association comes from the fact that psychedelic drugs trigger a non-ordinary mental state in which self-consciousness is diminished, allowing users to expand their perceptions and creativity. However, drug-induced altered mental states can be harmful and should not be attempted without careful medical supervision. Luckily, getting into a state of flow does not require using drugs.

Flow is technically defined as an optimal state where we feel and perform our best. We can get into flow through ordinary activities that require a high degree of focus, such as running a race or writing a book. In flow, we are completely involved in an activity for its own sake, not because we seek an extrinsic reward. Our sense of self diminishes, time passes imperceptibly, and our ability to tap into creative insights is heightened. The actions and thoughts follow one another, as in the assembly line of an imaginary idea factory. Surprisingly, flow seems to have long-lasting effects. Research done by Teresa Amabile shows that people who have experienced this state of mind report higher levels of productivity, creativity, and happiness for up to three days after experiencing a flow state.[39]

Advances in neuroscience have helped us better understand what is happening to our brain in a flow state. In flow, our brain waves move from the fast-moving beta waves characteristic of alertness when awake to slower-moving waves at the border of alpha and theta. Think about alpha as the state we are in when daydreaming. Theta is a little deeper— it manifests itself just before we fall asleep. Flow allows us to tap into deeply seated ideas in our subconscious mind that usually only show up when we are asleep or in a state of hypnosis, combining them with our conscious ability to create actionable thoughts. What happens when we tap into the vast amount of information accumulated over our entire lifetime, hidden at the subconscious level, and combine it with new sources of information available at the conscious level to achieve a desired goal is pure magic. Ideas pop into our heads out of nowhere, and creativity rises dramatically.

But there is more to the science of flow than just brain waves. A phenomenon called "transient hypofrontality," which refers to the temporary deactivation of the prefrontal cortex, allows the sense of self to disappear. This phenomenon has enormous implications for creativity. The dorsolateral prefrontal cortex is the brain area responsible for self-monitoring and impulse control. When this inner-critic voice goes quiet, we become more courageous and open to new possibilities, another creativity boost.

And it doesn't end there. Several neurochemicals that enhance pleasure and performance, including norepinephrine, dopamine, endorphins, anandamide, and serotonin, are released during flow, amplifying focus, increasing pattern recognition, and enhancing lateral thinking. In other words, the secret sauce that enables us to rise above the machines, giving us the edge needed to collaborate with them while staying vital in the automation economy, is readily available through flow.

So how do we tap into it? How do we get into a flow state to feel and perform our best? Let's unpack the fundamentals based on the current knowledge of this vast and evolving domain.

It starts with setting clear goals. Clear goals help focus our attention. The goal must be clear enough to identify precisely what needs to be done. This clarity improves concentration, increases motivation, and removes distractions. A flow-inducing goal needs to be challenging enough, just a little bit beyond our comfort zone, but not so much that it causes a breakdown. It should be stimulating enough to shorten our attention to the now but not so stressful to produce procrastination. Also critical is the magnitude of the goal. Big, challenging goals that take a long time to accomplish serve as a true north to guide our journey. But they need to be broken down into smaller, more manageable goals that can be completed at the moment. It's the difference between setting the goal to write a book, where you ultimately want to arrive over several months or years, and writing two pages in the next couple of hours, one paragraph at a time. Your full attention needs to be on the task at hand right now. If you shift your attention to the big, challenging goal, it may block flow.

Psychologists Robert Yerkes and John Dodson performed several experiments to understand the empirical relationship between stress and performance, resulting in what is known as the Yerkes-Dodson Law. In essence, it posits that performance increases with physiological or mental arousal, but only up to a certain point—what is known as the inverted-U model of arousal. The shape of the U curve varies based on the complexity and familiarity of the task. Complex tasks are best performed under a lower level of arousal, and simple tasks are best performed under high arousal levels.[40]

Flow seems to be triggered in what scientists call the flow channel—where the job is difficult enough to cause stretch but not so hard that it drives burnout. The trick with flow is knowing where that point is. Identifying this point becomes easier with practice. You want to be confident in your ability to perform the task, but it should make you sweat just a bit. You need to feel the stretch, acknowledge the struggle, and get enough physiological and mental arousal, knowing that what you are trying to accomplish is just a bit beyond your comfort level but definitely within reach. If it gets beyond that point and you feel like you want to give up, you are no longer in flow. This urge to give up indicates that you must adjust the difficulty level or take a break.

Other flow triggers include immediate feedback, autonomy, mastery, and what Steven Kotler, author of *The Art of Impossible*, refers to as the trinity of curiosity, passion, and purpose. Immediate feedback is about creating real-time feedback loops. This feedback is more accessible during certain activities, like sports, than in others, like writing. But not all triggers need to be activated for you to experience flow.

Autonomy is about being in charge of your thoughts and goals. When we have autonomy, we feel fully alive and in control, focusing our attention and concentrating on the task at hand, reducing anxiety and the fear of failure or irrelevance.

Mastery drives confidence. If you don't have any confidence, it is difficult to enter into a flow state because you quickly reach the point where the Yerkes-Dodson Law kicks in, lowering performance. But mastery does not need to be achieved all at once. You can build it over time with study and practice until you have enough confidence to stretch a bit beyond your comfort level, triggering flow.

The curiosity-passion-purpose triad works together to tighten focus by balancing human strengths and weaknesses. Curiosity is a powerful motivator and differentiator from machines. Passion may produce ego-driven focus, which activates the prefrontal cortex, preventing flow. But purpose acts as a counterforce, shifting the focus away from ourselves, achieving balance. These motivators can be stacked together to release dopamine and norepinephrine, generating flow.[41]

Paradoxically, flow is about gaining control by letting go of control. If you have ever tried to skateboard, you know what I am talking about. The fear of falling makes you tense, and your natural tendency is to stiffen your body, which is exactly the opposite of what you should be doing. You lose control if you feel tense and lock up your muscles. The same applies to flow. You cannot force flow. It needs to happen naturally. You can create an environment that facilitates flow triggers, but you can't schedule it and hope it will happen before your next appointment.

Remarkably, the most potent intrinsic motivational drivers—the secret sauce that differentiates humans from machines—are flow triggers. We have argued that curiosity, passion, purpose, autonomy, and mastery all play a role in achieving optimal human performance. If we learn to tap into these uniquely human capabilities, we will become more proficient in collaborating with the machines.

# The Magnificent Human Brain

The human brain is the most remarkable, effective, energy-efficient computing device ever created. AI has been trying to emulate what each of us can do in the blink of an eye, and it is nowhere near achieving the equivalent of general human intelligence. The more we understand how our brain works and how human intelligence contrasts with Artificial Intelligence, the greater our readiness to cooperate with AI, ask the right questions, and further evaluate AI responses that run counter to our intuition. In other words, knowing ourselves and understanding the mechanisms involved in our everyday thoughts, interactions, and decisions will increase our effectiveness in a world immersed in AI.

A purported human brain characteristic is that it contains two hemispheres that operate differently yet complementarily. Dr. Jill Bolte Taylor, a neuroscientist, had a unique research opportunity for which few brain scientists would volunteer. She suffered a massive stroke as a blood vessel exploded in the left half of her brain, allowing her to see the nature of the two hemispheres through her own brain. In the course of four hours, Dr. Taylor observed her brain functions—motion, speech, and self-awareness—shutting down step-by-step. She recounts her fascinating story in a TED Talk, where she explains that the two cerebral cortices are completely separate from one another, holding a real human brain for illustration purposes. She explains that our right hemisphere functions like a parallel processor in a computer while our left hemisphere works like a serial processor. The two sides communicate through the corpus callosum, which contains about three hundred million axonal fibers. "Because they process information differently, each of our hemispheres thinks about different things, they care about different things, and dare I say, they have very different personalities," she says.

She claims that our right hemisphere focuses on the present moment, learns kinesthetically through the movement of our bodies, and processes information from our sensory system simultaneously, allowing us to get a complete view of what it feels like in the present moment. On the other hand, our left hemisphere thinks linearly and methodically, picking up details that it categorizes and organizes, using the past as context and projecting into the future. It is responsible for the constant chatter in our brain reminding us of our obligations and giving us a sense of who we are separate from the external world—this is the capability Dr. Taylor lost on the morning of her stroke. She describes her peculiar experience: "I look down at my arm, and I realize that I can no longer define the boundaries of my body . . . the atoms and molecules of my arm blended with the atoms and molecules of the wall . . . and all I could detect was this energy."

She depicts the experience as a complete silence of her brain chatter, which allowed her to become captivated by the magnificence of the energy all around her. "I felt enormous and expansive. I felt at one with all the energy that was, and it was beautiful there . . . and I felt this sense of peacefulness . . . oh, I felt euphoria," she recalls. But her left hemisphere would step in once in a while and remind her that she needed help. She noticed her left arm was paralyzed and realized she was having a stroke. "Oh my gosh! I'm having a stroke! . . . Wow! This is so cool! . . . How many brain scientists have the opportunity to study their own brain from the inside out?" she said. Given her condition, it took her a while to get the help she needed, but fortunately, she fully recovered. She also benefitted from being able to recall and learn from the experience of having her left hemisphere temporarily disabled.[42]

Dr. Iain McGilchrist adds his perspective to the difference between the two hemispheres. In his book *The Matter with Things*, he explains that two conscious centers in our brain—the right and left hemispheres— produce different kinds of experiences. "To sum up a vastly complex matter in a phrase, the brain's left hemisphere is designed to help us apprehend, and thus manipulate the world, the right hemisphere to comprehend it, see it for all that it is," he writes.[43] He also explains that contrary to what was believed in the 1970s, both hemispheres are involved in reason, language, visual-spatial imagery, and emotion. The critical difference is the quality of attention—the left hemisphere has a narrow view of attention while the right hemisphere has this broad, open, vigilant attention. "The left hemisphere world is made up of fragments that are static, separate, decontextualized, disembodied, abstract, categorizable, and inanimate. Whereas in the right hemisphere, you see that things are all actually interconnected, that nothing is ever isolated completely, that they are also not fixed, uncertain," he explains.[44]

The human brain is truly a remarkable engineering feat. However, despite its impressive capabilities, it does have some limitations. One of our most pronounced limitations is our limited working memory—the equivalent of Random Access Memory (RAM) in computers. A critical difference between our brain's capabilities and those of computers is that we can upgrade our computer's RAM or buy a more powerful computer when the installed RAM reaches its limitations. However, the human brain, at least for now, is not upgradeable.

Harvard cognitive psychologist George Miller published one of psychology's most famous papers, "The Magical Number Seven, Plus or Minus Two: Some Limits on Our Capacity for Processing Information." In this paper published in 1956, Miller proposed as a law of human cognition and information processing that humans can effectively process no more than seven units, or chunks, of information, plus or minus two pieces of information, at any given time. That limit applies to short-term memory and many other cognitive processes, such as distinguishing different sound tones and perceiving objects at a glance. Do you ever wonder why our phone numbers contain seven digits excluding the area code? This limitation was taken into consideration at a time before smartphones when we used to memorize people's phone numbers and primarily make local calls.[45]

In several follow-up studies, researchers found that we can usually hold about three or four items in memory simultaneously. This limitation is why "the rule of three" is used extensively in writing, speaking, music, and marketing. Ideas presented in threes are easier for the human brain to absorb and are, therefore, more memorable. Steve Jobs, a master in communications and marketing, was an avid user of the rule of three. You can observe how he often used this rule in his presentations as he referred to products as "thinner, lighter, and faster."

But here is where it gets interesting. Modern science is discovering that the human brain is much more neuroplastic than we realized. With practice, we can train our brain to perform wonders that at first may seem impossible, like holding much more than just three, four, or seven pieces of information at once. Take, for instance, Marc Lang, a player of what is known as "blind rapid chess." He never gets to see his opponents' boards. For each opponent, he is only told their latest move. Everything else needs to be stored in his memory. He plays twelve opponents simultaneously, meaning he has to hold hundreds of positions and moves in his mind concurrently, defying George Miller's and fellow researchers' studies by orders of magnitude. Here is how Marc describes his experience: "There is kind of a new room opening up in my mind, and inside of this room, there are all the boards. There is all dark, and I say, ok, show me board eleven. And board eleven is coming out of the dark, and I can see it." Marc doesn't know how he does it. He says the games stick to his mind without him doing anything particular. Marc is just an ordinary guy, proving that we can do extraordinary things with our brains.

Marc's story was told by Todd Sampson, an Australian television personality, in a series called *Redesign My Brain*, where Sampson puts brain training to the test as he undergoes a radical brain makeover to prove that any brain can turn into a super brain.[46] In the first episode, Sampson trains to expand his mind power to compete in some of the most challenging brain competitions on the planet. Dr. Michael Merzenich, Chief Scientific Officer at Posit Science, Professor Emeritus at UCSF, and a pioneer of the brain plasticity revolution, is Sampson's mentor. First, Sampson goes through a battery of tests and measurements to capture a baseline of his current brain state, recording its speed and alertness. The first test and exercise series aim to increase his recognition, or thinking speed. The baseline is measured by capturing the accuracy and reaction times to objects that flash on his screen.

Our thinking speed naturally declines as we age. For example, teenagers can sample information seven times per second, while an eighty-year-old does so only twice per second. The good news is that, with practice, thinking speed can be improved at any age. There are several apps, like Luminosity, that you can use to enhance your thinking speed. Surprisingly, one of the best activities to increase thinking speed is juggling. By learning and practicing it daily for a few minutes, you can significantly boost your ability to think quickly and improve your attention and focus.

Next, Sampson focuses on improving his attention or alertness. If you have ever been stunned by a magic trick, most likely the magician was taking advantage of a human weakness, what neurologists call the mind blink, a state in which the brain does not see what is right in front of our eyes. Magicians are masters at manipulating human attention, awareness, cognitive processes, and even emotions. This manipulation is what magic is all about. A favorite magician trick is to introduce humor in the middle of a performance. While our brain is perceiving humor, magicians can perform actions that are perceptually invisible to our brain. They are taking advantage of the fact that the brain needs dedicated focus to pay attention, and distractions keep us from experiencing full awareness. That is why multitasking is a fallacy that keeps us from doing our best work.

The challenge we all face in the modern world is that distractions surround us. Every ding from our smartphone reduces our attention span and focus. We are training our brains to do just the opposite of what could improve our alertness. Even worse, these distractions can be addictive. Our brain craves looking at the alert, the new email, or the social media feed because they represent novelty. Every time we stop what we are doing to look at the message that just came into our smartphone, a small dose of dopamine gets released into our bloodstream. Dopamine is very pleasurable but can also be addictive. If our dopamine level drops below a baseline, we become deeply unmotivated. Additionally, as we saw in the discussion about flow, dopamine is one of the neurochemicals that support optimal performance. But short dopamine spikes from indulging in activities such as checking our phones are followed by a dopamine crash that keeps us below a baseline for an extended period.

In her book *Dopamine Nation*, Dr. Anna Lembke, Chief of the Stanford Addiction Medicine Dual Diagnosis Clinic at Stanford University, explains that the human body contains mechanisms to maintain homeostasis. That means it is always trying to get to the point of equilibrium. This principle also applies to how our brain deals with the pain-versus-pleasure equation. When there is a spike in pleasurable activities, like checking our phones or eating ice cream, the brain creates pain-inducing reactions to maintain equilibrium. According to Dr. Lembke, this explains why in developed countries where we have abundant access to constant pleasure, we experience the highest rate of suicide, depression, and physical pain. She states that there are only two ways to keep this equation in balance: temporarily pursuing pain or abstaining from pleasure. "Pressing on the pain side of the lever can lead to its opposite—pleasure," she says.[47]

The dopamine released from pain appears to be more enduring and can have healing effects. For example, an ice bath will send shock waves through our body, followed by a gradual 200 percent rise in dopamine, the equivalent of the increase in dopamine people experience when snorting cocaine. Exercise is another effective practice, gradually increasing dopamine by 100–200 percent. Abstinence, or intermittent dopamine fasting, is also effective in maintaining equilibrium. Refraining from foods you crave or taking a break from digital interactions, like checking your phone regularly or Netflix binging, helps the brain reset its dopamine levels. But if you have developed an addiction and suffer from compulsive behaviors, the abstinence needs to be extended to thirty days. This rest gives the brain enough time to reset its reward circuits, allowing withdrawal symptoms to subside.

Back to Todd Sampson: as he continued to work on his alertness, he also worked on improving his peripheral vision, which degrades as we age. By the time we reach our sixtieth birthday, we have lost about one-quarter of our peripheral vision, and by the time we are eighty, we can no longer see about half of what is in front of us. Once again, the good news is that we can see dramatic improvement with practice, regaining about twenty years of peripheral vision capability after practicing for a few hours. Lastly, Sampson learned how to use mnemonics to help him improve his working memory. To help him memorize a deck of cards, he enlisted the help of Tansel Ali, an Australian Memory Champion who has memorized two entire Yellow Pages books.

After four weeks of training, Sampson improved his reaction time, or thinking speed, from 0.893 to 0.493 seconds (almost twice as fast) and increased his accuracy from 62.0 percent to 98.3 percent. He also made tremendous progress with his alertness, going from a blink accuracy of 64.4 percent to 95.6 percent, a near-perfect score, indicating that his brain was operating in a high alertness state, responding as fast as humanly possible. After his coaching lessons from Tansel Ali and three weeks of practice, Sampson also participated in the World Memory Championship and, surprisingly, recalled all fifty-two cards shuffled in a deck with 100 percent accuracy, awarding him the honor of ranking fifty-seventh in the world for card memorization.

The most important lesson from Sampson's experience is how quickly and powerfully we can change our brain and improve its capabilities. After only one month of training, his thinking speed nearly doubled, his attention became much more focused, and his memory improved radically.

In a world where we need to work side by side with incredibly capable machines endowed with Artificial Intelligence, we must take good care of the body and brain that give us human intelligence, to stay competitive. The goal is not to compete with the machines head-on. With some training, we can expand our memory capabilities and learn to recall an entire deck of cards, but the machines can do the same and much more in thousandths of a second. The point is to keep our mental capacities sharp, and to keep our mind and body in harmonious, healthy equilibrium, so that we can use our human intuition and our unmatchable—at least so far—general intelligence to complement the machines' capabilities.

Before we conclude our discussion of the human brain, I would be remiss if I did not mention the work of psychologist and Nobel Prize winner Daniel Kahneman, who has made valuable contributions to our understanding of our decision-making process. In his best-selling book *Thinking Fast and Slow*, we learn about what Kahneman calls Systems 1 and 2. In essence, System 1 is our automatic, intuitive response system, responsible for the things we do every day without much thought or effort, like putting one foot in front of the other when we walk. This fast-thinking process was fundamental to our survival when faced with dangers in the early days of our history. System 2 is the thoughtful and deliberate process that allows us to perform complex tasks like making calculations. It is a much slower process requiring thoughtful consideration of the many factors typically involved in making difficult decisions.[48]

Kahneman's work helps us understand why we do the things we do. We often think we are taking certain actions for rational reasons using System 2 when in reality, we are just in System 1 reactionary mode. This reactionary, emotionally charged decision-making process often leads to suboptimal decisions unless we have been trained extensively in making those fast decisions in the pertinent domain. It also leads to what Kahneman calls "cognitive illusions," making an erroneous decision that "feels" right, even when it is not rational. An example would be falling for a charming and convincing scammer who gains our trust and persuades us to make a poor investment decision, one that we have been warned against and would not usually make. Cognitive illusion is what allowed Bernie Madoff to get away with creating one of the largest Ponzi schemes in history.

When we are in what Kahneman calls cognitive ease, meaning when System 1 is running the show, we tend to be more impulsive, emotional, and optimistic, following our intuition. Understanding this concept can be very helpful in collaborating with AI. AI can make System 2 decisions much faster than we can, which can be extraordinarily helpful in many circumstances. However, humans can use intuition, empathy, and emotional intelligence, which the machines lack, resulting in a symbiotic relationship.

## Social and Emotional Needs

Researchers have long tried to understand what makes humans happy. Dr. Robert Waldinger, a professor of psychiatry at Harvard Medical School at Massachusetts General Hospital and coauthor of *The Good Life*, explains how we mistakenly underestimate how vital relationships are for our well-being. "We seem particularly bad at forecasting the benefits of relationships," he writes with his coauthor, Marc Schulz. "A big part of this is the obvious fact that relationships can be messy and unpredictable. This messiness is some of what prompts many of us to prefer being alone." Indeed, human relationships can be messy, but who is better equipped to navigate the delicate balance of keeping relationships healthy than human beings? AI does not feel anything, so it can't be very helpful in empathizing with human beings' emotional and social needs.

In their book, Waldinger and Schulz reference a study that began in 1938 and followed the lives of 724 Harvard students as well as low-income boys from Boston. According to the researchers, this is the world's longest scientific study of happiness to date. The study, which is ongoing, has expanded to include the original participants' spouses and children, consisting of over two thousand people. At about fifteen-year intervals, the researchers interviewed the participants to uncover the key to happiness. The results indicated that strong relationships were the most accurate predictor of people's happiness throughout their lives. They are "intrinsic to everything we do and everything we are," the authors said.[49]

Humans have an inborn need to connect with other humans. This need is the cornerstone of our social fabric. If you recall our earlier discussion about Maslow's hierarchy of needs, love and belonging come right after our physiological and safety needs. The need to love and feel loved is a universal trait that allows us to feel compassion and to understand and care for each other. It starts with our first experience as we come into this world and feel our mother's heartbeat and caring touch. We first learn how to love within our families, the foundational social structure for human beings.

In the United States, nearly twenty-four million children live in a single-parent family. The social impact is significant: Nearly 30 percent of single parents live in poverty compared to just 6 percent of married couples. The emotional and behavioral effects on children are substantial. Kids from single-parent families are more likely to face emotional and behavioral health challenges—like aggression or engaging in high-risk behaviors—when compared to peers raised by dual parents.[50] When the family structure starts to falter, the greater society suffers the consequences, ranging from low developmental outcomes to drug abuse and crime.

Beyond the family structure, our ability to relate to and effectively communicate and cooperate with other humans is fundamental to our growth and development. We have discussed the drivers of optimal performance from an individual standpoint, but the reality is that our performance is bounded. To accomplish bigger goals, we need to work with others, and our success is often determined by our ability to network professionally and socially. The best networks are the ones that bring together people who are naturally good networkers. They are connectors eager to use their relationships to support others and the causes they care about. By bringing them together, a network of networks is created, amplifying the power and reach of those relationships. According to Albert-László Barabási, Professor of Network Science at Northeastern University and author of the book *The Formula*, there are specific universal laws of success. The first law states that the harder it is to measure performance, the less performance matters. What he means is that unless there are precise parameters performance can be measured by—for instance, how long it takes a competitive runner to complete a 100-meter dash—the performance itself is not a key determinant of success. "No matter the field, discipline, or industry, if we want to succeed, we must master the networks," he writes.[51]

Human uniqueness lies at the intersection of chemically induced, electrically charged brain activities, social interactions, and spiritual influences. The latter are more nuanced and require a more profound discussion that we will explore at the end of this book. By focusing on the capabilities that are uniquely human and developing the skills and social connections that the machines cannot duplicate, we will be better prepared to participate in the AI-driven automation economy. We will look at how to develop these skills in the next chapter.

## Takeaways

» Science struggles with the definition of consciousness, so we are baf-fled by questions about machines becoming sentient since we don't understand what that means.

» The advancement of AI is forcing us to ask profound questions that philosophers have wrestled with for millennia, such as "Who are we?" How are humans different from a machine that could surpass our intellect?

» Machines don't have desires distinct from the objective functions imparted to them by humans. On the other hand, humans are moti-vated by a complex set of factors involving biology, the environment, and spiritual matters, although science may contend the latter.

» Understanding our motivations and how the human brain works is paramount to the future of work as we prepare to collaborate with intelligent machines.

» Neuroscience is advancing our understanding of neuroplasticity and flow, helping us adopt practices that improve the brain and increase performance.

# CHAPTER 5:

# Learning the Skills of the Future

No other technology is expected to be more impactful to knowledge workers than AI. It is estimated that AI will increase the productivity of knowledge workers more than fourfold by 2030. This increase represents higher global labor productivity of $200 trillion, dwarfing the $32 trillion in total knowledge workers' salaries. AI training cost is expected to continue declining at an annual rate of 70 percent. The amount of time to complete coding tasks with AI's assistance today is half the amount a software engineer can achieve without AI, based on current technology. The time and cost for AI to create a graphic design are estimated to cost $0.08 in less than a minute, compared to $150 for five hours of human labor.[52] These are just a few examples of how AI will revolutionize the world of work.

The environment for knowledge workers is changing right in front of our eyes. This astounding increase in productivity means that companies can produce much more with fewer people. The lucky few who get to keep their jobs will be highly skilled in using AI effectively, combining their human analytical and intuitive capabilities with the machines' data-crunching power to produce goods much more efficiently. The ability to collaborate with AI will be one of the most sought-after skills in the future. Learning to do this effectively is paramount to staying competitive in the labor market.

In an era when technology is advancing at exponential rates, it is very difficult, if not impossible, to predict what future jobs will look like in ten, twenty, or thirty years. But we can look at current trends, analyze projected demand in the near future, and try to extrapolate the general direction in which jobs are moving.

The *Future of Jobs Report 2020*, published by the World Economic Forum (WEF), provides a glimpse into job roles that are increasing in demand and those that are decreasing. It is not surprising to see roles such as data analysts and scientists, AI and machine learning specialists, and process automation specialists on top of the increasing demand list. Positions such as data entry clerks and administrative and executive secretaries are on top of the decreasing demand list, as expected. However, what is insightful is the increasing demand for roles that require broader, well-rounded soft skills, such as strategic advisors and organizational development specialists.

The report also details the top fifteen skills for 2025. Again, it is not surprising to see the types of skills that one would expect in a technology-dominated world, such as critical thinking and analysis and technology design and programming. What is informative is the increasing prominence of soft skills, such as creativity and originality. Also noticeable are initiative, resilience, stress tolerance, flexibility, and emotional intelligence.[53]

The data may be signaling that individuals with the unique ability to combine hard and soft skills will be in high demand in the future. The trend is pointing toward the need for workers to bring the total capacity of the human brain to the job, indicating that well-rounded individuals might be more adaptable and, therefore, more valuable than those who are more narrowly focused.

This emerging trend corroborates David Epstein's contention that generalists will do well in an environment where AI can do the specialized work. In his extraordinary book, *Range: Why Generalists Triumph in a Specialized World*, Epstein provides compelling evidence that our unique ability to see the big picture and integrate broadly is and will likely continue to be, in the foreseeable future, a competitive advantage against the machines. He quotes Gary Marcus, the psychologist and neuroscience professor mentioned earlier: "In narrow enough worlds, humans may not have much to contribute much longer. In more open-ended games, I think they certainly will. Not just games, in open-ended real-world problems we're still crushing the machines."[54]

The future will require people with the ability to solve wicked problems that are complex, constantly changing, and for which there are no prescribed formulas. In a complex, interconnected, and fast-changing world, we need to develop a workforce that is proficient at lateral thinking—the ability to search for input from multiple domains and connect the dots in unique ways to gain new insights. Tasks that are restrained and repetitive are more likely to be automated. To compete and collaborate with the machines, humans must use conceptual reasoning skills to connect new ideas across contexts and apply conceptual knowledge from one problem domain to an entirely new one.

Management will need to take these considerations into account as it plans for organizational development. Hiring processes may have to evolve from seeking candidates who can fill specific roles based on specialized knowledge and experience to a less restrictive practice where companies seek to hire people who can demonstrate the ability to fill multiple undefined roles. Diverse experiences may become more valuable than specialized knowledge. Training programs will need to adapt so that people can be given the opportunity to be exposed to a broad set of principles and experiences.

One challenge we face today is that we are educating the future workforce too narrowly. The education system needs to prepare students to deal with complexity better, use abstraction and decomposition to attack complex problems, and clearly define problems before trying to solve them. James Flynn, a political studies professor at the University of Otago in New Zealand, states, "Students need to be taught to think before being taught what to think about." In a world where we can easily find answers through searches and the help of AI, it is substantially more important to know what questions to ask than what answers to give.

The unrelenting advances in AI will impact employees and independent professionals equally. As companies seek to reduce their dependence on human labor, we will continue to see more independent professionals participate in the labor market—commonly known as the gig economy. These people work as freelancers or entrepreneurs. New platforms facilitate the matching of supply and demand, and communication tools such as videoconferencing, file-sharing, and text messaging allow workers to collaborate, freeing freelancers from any geography-based boundaries. An increased supply of independent professionals and rising AI-assisted productivity will pressure labor rates. Competitive forces will drive many of these workers out of the market, except those highly capable of leveraging AI to compete.

The implications for organizational development are enormous. Companies will likely evolve to look more like the types of organizations described by Salim Ismail in his book *Exponential Organizations*. They typically contain just a few employees who are highly effective in using technology to produce extraordinary results. These organizations balance left-brain capabilities such as order, control, and stability with right-brain proficiencies in creativity, growth, and uncertainty.[55]

# Generative AI

Generative AI has taken the world by storm, awakening the general public's interest in interfacing with AI tools that can help them become more productive. There is no question that Generative AI can be a huge productivity enhancer, but using it effectively requires applying new skills that most people may not be familiar with.

In practical terms, we can start interacting with AI and improving the quality of the collaboration immediately. By learning some fundamental skills and practicing them, we can quickly become better at knowing what questions to ask and how to ask them. There is no better way to do this than to take advantage of the Generative AI tools that have recently become available to the general public.

For example, ChatGPT, the most popular Generative AI tool to date, is based on a Large Language Model (LLM). A language model synthesizes sentences from the enormous dataset it was trained on. In the case of ChatGPT, it has allegedly been trained on several petabytes (a million gigabytes) of data, including the Web, books, articles, and many other data sources. The way ChatGPT operates is based on probabilities—it is a sophisticated prediction machine. In simple terms, the tool maps the probability of the next word or sentence based on what came before or its context. ChatGPT and other LLMs like it are improving so rapidly and producing such extraordinary results that sometimes it may feel like they have humanlike intelligence. However, it is essential to keep in mind that AI is not conscious. It is simply remarkably good at simulating cognitive proficiency.

An important consideration for working effectively with Generative AI is knowing that the tool holds no actual understanding of the input provided. The more context it has to work with, the higher the probability that the outcome will be helpful. Therefore, the prompting to obtain the desired output is critical in interacting with the tool.

The more context you provide, the more likely you will get a richer, more meaningful output. For example, you can give the tool a role you want it to adopt, such as a travel agent, journalist, consultant, or any other role that fits your request. Additionally, you can prompt the domain or knowledge realm that should be used, such as methodologies, philosophies, etc. You can also tell the model what inputs you will provide and specify what kind of outputs you expect back. For example, are you looking for a list of items, or do you need the result in narrative form? What voice should be used in the narrative, that of an expert making a formal presentation or a layperson having a casual conversation? The more specific the prompting, the better the outcomes.

Another way to improve your prompts is to "pump" the machine, meaning that you provide examples to help AI come up with more. Let's say, for example, that you have a list of possible titles for an article you plan on writing. Provide the list to AI and ask it to come up with alternative titles. By providing examples that AI can use as a starting point, the output will be more aligned with your ideas, and you won't waste time dealing with results that are way off base. You can also continuously improve the output by taking the initial response, refining it to your liking, and providing it as the following prompt. By repeating this, you can truly collaborate with the tool to derive your desired outcome. The ability to prompt AI is such a valuable skill that new jobs are beginning to emerge. For example, Boston Children's Hospital posted a job for an AI Prompt Engineer—someone trained to use ChatGPT to extract the most helpful results—as reported by *Good Morning America*.[56]

Keep in mind that a Generative AI tool can only provide an output based on the data it has been trained on. For example, ChatGPT was trained on information that ended in 2021. Developments that have occurred since then will not be considered in the output. It is also possible that Generative AI will create completely inaccurate outcomes. The machine cannot know what is accurate as it develops answers by probabilities. Another possibility is that the model will be biased, but the bias can be somewhat reduced if the prompts are chosen carefully.

In the case of creative outputs, such as images and videos, users can prompt characteristics such as composition and style. These tools allow humans to elevate creativity to an otherwise unattainable level. The AI itself is not being creative. It is simply mashing pieces together according to human prompts. In most cases, it will not provide a final output, but will get the user to the finish line much faster than without AI's help. The power of collaboration is that AI can help humans become more creative by very quickly expanding the realm of what is possible, providing an infinite number of new combinations that expands our imagination.

Everything Generative AI produces is a regurgitation of content created by someone else, so human intuition becomes essential to review and refine its output. The quality of the cooperation with Generative AI increases manifold by analyzing the results, improving the questions, changing the prompting, and continuously reiterating. When users treat AI like a collaborative partner, their capacity expands and their performance increases.

Imagine having your own AI assistant you can call upon on demand, anytime, anywhere. This AI assistant is trained on vast knowledge based on the Web, books, articles, and, most importantly, data about you. It knows you better than you do yourself, including your habits, abilities, knowledge, physical characteristics, personality, best learning modes, moods, and more. AI assistants can potentially accelerate human learning to unprecedented levels in the future. Imagine having this all-knowledgeable tutor at your disposal, ready to help with whatever knowledge you need—that is where the technology is headed.

Educators are alarmed by ChatGPT, fearing that students will use it to cheat on standardized testing and to complete homework assignments. New York City prohibited using the tool across all devices and networks in New York's public schools. Jenna Lyle, a department spokesperson, justified the decision due to "concerns about negative impacts on student learning and concerns regarding the safety and accuracy of contents."[57] While some school districts are prohibiting the use of ChatGPT, Steven Mintz, Professor of History at the University of Texas, is encouraging his students to learn how to collaborate with the tool. He believes it will strengthen students' writing abilities despite its shortcomings. "The next task for higher education, then, is to prepare graduates to make the most effective use of the new tools and to rise above and go beyond their limitations. That means pedagogies that emphasize active and experiential learning, that show students how to take advantage of these new technologies and that produce graduates who can do those things that the tools can't," he writes.[58]

Understandably, educators are concerned about a new tool that may not provide accurate content. However, this is the future, and the sooner we embrace it, the better prepared our youth will be for what is to come, which will be orders of magnitude more powerful and pervasive. We can't keep the students in a bubble while the entire world is leveraging Generative AI to create content, enhance creativity, and find answers more quickly. As previously discussed, the same type of resistance occurred a few decades ago when electronic calculators became available. Educators missed the point then that calculators would become so pervasive that no one would ever need to do calculations by hand again. They are missing the point again, this time with AI.

We have seen the dawn of digital natives who became very proficient using digital tools, leaving the older generations behind in their ability to be effective contributors. We are now seeing the emergence of the AI natives' generation, and its progression will be much faster. Students and professionals unable to collaborate effectively with AI will be left behind. We can't underestimate the magnitude and impact of this change. Instead of prohibiting it, we must encourage the current generation to explore ChatGPT and all the new AI tools sprouting at a remarkable pace. Educators have an essential role in teaching the limitations and potential dangers of the tools. Guardrails must be put in place to protect against inappropriate exposure and usage. Still, we cannot be naïve about the unstoppable proliferation of this technology. The role of educational institutions must shift to enable the future generation of AI natives to effectively leverage their human intuition in collaboration with AI—more on that later in the chapter.

## Learning to Learn

As technologies advance and influence all modes of operation and production, we must adopt a lifelong learning philosophy. The obsolescence of knowledge will continue to accelerate, so we must adapt and embrace a new form of learning that will prepare us to be more agile and more open to acquiring new skills.

Learning a new subject can bring conflicting feelings of excitement and discomfort. At first, curiosity provides the fuel needed to start the exploration process, and our natural yearning to learn something new creates anticipation. But as we dive into it, entering an unfamiliar territory where we may not even understand the terminology can be disheartening.

As a beginner, it is crucial not to get frustrated and discouraged. Being aware of this initial difficulty can help release the anxiety.

The first step in becoming knowledgeable in a new subject is to gain familiarity with the terminology being used. Many domains contain their own language—the fundamental means of human learning—so vocabulary acquisition is essential. Reading related articles and books can broaden our vocabulary so we can gradually understand the subject better. The key is not to get frustrated and to keep reading. At an unconscious level, the brain will be seeking pattern recognition, and before we realize it, bits of information will be put together, and comprehension will begin to emerge.

The same concept applies to learning a new physical skill. It is impressive how young children can quickly learn how to ride a bike or downhill ski. Once those skills are acquired, we build what is referred to as "muscle memory," so riding a bike becomes second nature. As we age, developing new physical skills becomes more demanding and requires more time and practice. However, as discussed previously, science is discovering that the human brain is gifted with more neuroplasticity than we may have realized.

Muscle memory is excellent for performing increasingly difficult physical tasks. Musicians spend considerable time practicing scales to play the notes in the scale effortlessly. This ability removes cognitive load during performance, allowing the musician to focus on more creative aspects of their playing. The techniques built into muscle memory must be refined because once they are set in place, it becomes very hard to unlearn them.

Take, for example, learning how to ride the "backwards bicycle." The backwards bicycle is a modified regular bike which turns the front wheel in the opposite direction of the handlebar. If you turn the handlebar to the right, the bike goes to the left, and vice versa. What is fascinating about the backwards bicycle is that it looks deceitfully simple to ride. Most people think they can easily ride it since they already know how to ride a regular bike and conceptually understand that they need to turn to the right to go to the left. What actually happens when people hop on the backwards bicycle is insightful and entertaining—I would encourage you to watch the YouTube video included in the references.[59] Riding the backwards bicycle requires unlearning previously acquired muscle memory, which takes more time and effort than is apparent.

We learn from the backwards bicycle experiment that neuroplasticity allows us to gain new skills with deliberate practice, creating new neural pathways while weakening others. But to learn, we must let go of our fears and overcome the immediate perception that the task is impossible. We must also accept the difficulty in acquiring new skills, especially when opposing habits have been ingrained in our brains. This acceptance requires humility and a positive attitude. Finally, we learn that conceptually knowing something does not readily translate into understanding it, and understanding it does not automatically turn into doing it. Nobody has ever become a top sports player by reading a book.

The backwards bicycle experiment provides another critical insight. Sometimes to grow and learn new skills, we need to unlearn old habits and let go of deep-rooted dogmas that no longer serve us. Aidan McCullen, the host of *The Innovation Show*, tells us in his newsletter about a process called apoptosis in which the human body gets rid of fifty to seventy billion unwanted cells per day so that the body can stay healthy. He uses this example to draw a parallel to how successful companies like Apple and Amazon relentlessly renew their businesses by cannibalizing old ones. He quotes Jeff Bezos, who moved Steve Kessel, the executive responsible for Amazon's paper distribution business, to run Kindle: "Your job is to kill your own business . . . I want you to proceed as if your goal is to put everyone selling physical books out of a job."[60] We can use the apoptosis metaphor for growth and development as well. Through training, we can strengthen new neural pathways and create new habits while atrophying old ones that are no longer serving us.

In his book *Talent Is Overrated*, Geoff Colvin tells us that a significant amount of research on performance science has accumulated over the last three decades, pointing to the effectiveness of deliberate practice in acquiring new skills. He defines deliberate practice as activities specifically designed to improve a particular skill. These activities require a great deal of repetition and mental effort and are most effective under the supervision of a teacher or coach who provides constant feedback.

Colvin argues that outstanding performance is not due to natural gifts. It is, instead, the result of many hours of dedicated effort. "Research suggests very strongly that the link between intelligence and high achievement isn't nearly as powerful as we commonly suppose," he writes. He suggests that many years of deliberate practice change the body and the brain.[61]

Colvin's work corroborates the general idea of brain plasticity discussed in the show *Redesign My Brain* with Todd Sampson and the backwards bicycle experiment. The key takeaway is that human intelligence is a multidimensional, fluid concept that can be reshaped—we are not constrained by the genes we inherited from our parents. We must approach the learning process by letting go of what we can't control—our natural abilities, our past, and our environment—and embracing the factors that we can influence—our attitude, effort, and patience. Studies have shown that it takes, on average, thirty days to develop a new habit. We tend to overestimate our ability to accomplish things in the short term and underestimate our long-term achievements. We can't expect to change what we know and do overnight—learning takes time.

We should also recognize that when faced with a significant learning challenge, we tend to become paralyzed by complexity. We should not let that stop us from moving forward. Many subjects can seem complex and difficult until we understand them. Understanding requires breaking the complexity into smaller chunks that are easier to absorb. If we ever become overwhelmed by the complexity of the subject at hand, we need to take a step back and start with something simpler. Before we know it, the pieces will begin to come together, and what seemed unattainable initially will become easier to learn. The most important thing is to lean into action.

As discussed at the beginning of the chapter, certain roles, such as data analysts, scientists, and machine learning specialists, will be in high demand in the future. These roles will require hard analytical skills that can be learned if we are willing to put in the long hours of practice and dedication required. Other high-demand roles, such as strategic advisors or organizational development specialists, will require soft well-rounded skills that need a different learning approach. In this case, instead of specializing in a particular domain, the goal is to become a polymath.

Some experts suggest that the deliberate practice purported by Colvin is only effective for improving certain skills that are very specific and that follow a prescriptive formula. To address more complex problems that require the ability to look across domains to find connections or common patterns—known as lateral thinking—a broader, more generalized educational approach may produce better results.

Becoming a polymath requires reading broadly across many subjects, gaining exposure to many different cultures and philosophies, and seeking a broader understanding of the world. This broad exposure helps us develop empathy and connect the dots across domains in unique ways. We are not talking about becoming a jack of all trades and master of none. We are talking about developing a certain depth level in a few subject areas but then aggressively expanding the scope of our learning to many different domains.

Harvard Professor Robert Kegan, a pioneer in the study of adult development and learning, introduces the concept of the subject-object shift in his book *In Over Our Heads*. The idea is to move what we know from "subject"—something controlling us—to "object"—something we can control. The overall premise is that the more in our lives we take in as an object, the more clearly we can see the world, ourselves, and the people in it. He suggests that becoming an adult isn't about learning new things or adding things to the "container" of the mind. It's about transformation—changing how we know and understand the world. He prescribes a learning environment that is much more friendly to contradictions, oppositeness, and being able to hold onto multiple systems of thinking.[62] This approach to learning is particularly valuable as we prepare to enter a world that is increasingly uncertain and ambiguous.

The challenge we face as we prepare for the skills of the future is that we don't really know what the future holds. We can take an educated guess that AI will be a dominant force, so the more we learn about how to collaborate with AI, the better. As stated earlier, we can start practicing that now by becoming familiar with the many Generative AI tools that have become available recently. We can also choose one or two areas where we want to gain some in-depth knowledge, preferably leveraging analytical skills, as they will likely be helpful in any interaction with machines. Beyond these specific areas, we ought to go broad, embracing diversity.

We should read as widely as possible. Ideally, we should read at least one book a month, possibly more. We also need to instill the discipline of reading about subjects we are entirely unfamiliar with. When traveling abroad, instead of being a tourist, we should adopt a scholar's mindset and try to absorb as much of the local culture as possible by spending time with the locals, visiting their homes, eating their food, and observing their practices. We must train our brains to appreciate multiple perspectives, know what incisive questions to ask, listen intently without judgment, and gain empathy for other human beings.

Our brain is a pattern recognition machine. Have you ever had the experience of reading a book or having an intellectual conversation with someone and finding this stimulus triggering new thoughts and ideas? The more diverse content we feed into our brain that it can use as context, the more it can make connections, find new patterns, and increase our overall comprehension of the world around us. Putting all the pieces together is believed to be the primary role of the brain's right hemisphere.

Dr. Iain McGilchrist reminds us of the importance of the right hemisphere of the human brain. He tells us that the left hemisphere helps us get a grip on the world so we can get what we need from it, while the right hemisphere helps us comprehend it and see it for all it is. Dr. McGilchrist's work is very pertinent to the sweeping premises of this book—that machines will do the work that gets the things we need so that humans can spend more time contemplating the more significant, broader aspects of life. Based on his assertions, we can reason that the brain's right hemisphere will be much more influential in the future. Therefore, it would make sense to focus our development efforts on right hemisphere matters. It is not that we should ignore the left brain—it still performs an essential function—but we should take a broader approach to our learning to expand the right brain's fundamental role in helping us see the big picture. "Unfortunately, we now are paying so much attention to what the left hemisphere tells us in its very technical way that we think we live in this map rather than in the real living world. And I think that's where we could begin thinking about what is the matter with us these days," he says.[63]

# Education Disrupted

Education is primed for disruption and will likely see significant changes over the next decades. Our education system was built on a model designed for the Industrial Revolution and hasn't changed much since. The ethos of the current education system can be attributed to Johann Gottlieb Fichte, a German philosopher who said, "Education should aim at destroying free will so that after pupils are thus schooled, they will be incapable throughout the rest of their lives of thinking or acting otherwise than as their school masters would have wished."[64] The industrialists who needed robot-like human labor liked this idea, resulting in what we refer to today in the United States as high school. This model brought many benefits, like standardization, literacy, and math, but it is no longer helpful in a digitized world where we need to be proficient at collaborating with AI.

The problem is that academia is very traditional and resistant to change, so progress has been slow in a world that is moving very fast. In an environment resistant to change, you often need a crisis to overcome the prevailing system. The pandemic was the crisis that served as a catalyst to overcome one type of resistance in the education system—it has forced schools to rethink the way education is conveyed and has opened the doors to a hybrid approach that combines digital and in-person forms of content delivery.

The digital technologies we are using today to connect online and experience remote delivery have been around for quite some time. But before the pandemic, with few exceptions, professors were very reluctant to use this remote delivery model. We are now beginning to realize that a hybrid delivery can be more effective than the traditional in-person-only delivery model.

Higher education has been on an unsustainable path of cost spiraling out of control, resulting in one of the most serious financial concerns in the United States—the student debt crisis. Now, universities are being forced to rethink their cost structure. People are comparing the education they can get online from low-cost sources with the education they are getting from traditional universities and questioning the value they are getting for their money. This competitive force will drive reduced enrollment, consolidation, and the closing of many small private colleges that will not survive in this environment.

In the United States, proposals are being considered for government-funded student debt forgiveness, creating controversial political debates. Notwithstanding the merit of such proposals, debt forgiveness does not address the root cause of escalating higher education costs. A possible solution is to let technology-driven deflationary forces play their role, as elaborated below.

The current education model needs to be turned entirely on its head. Things that are centralized today, like curriculum development and credentialing, would benefit from a decentralized model. And things that are decentralized today, like content delivery, would be much more effective if centralized. With the help of AI, a single teacher can effectively deliver content to thousands of students across the globe, reducing costs.

We will likely see the development of a new model for curriculum development and credentialing based on decentralized networks and modularized components that can be customized and put together into unique curriculums. In this model, students could take classes from multiple institutions and combine them into certificates, nanodegrees, and full degrees. Under this structure, students would not need to be resident students of one institution—they could earn their degree concurrently with work experience throughout their careers.

One of the potential challenges to this model is that we will need new innovative mechanisms for developing curriculums that receive accreditation by combining credentials from multiple courses and experiences. The solution to this challenge may be based on Blockchain technologies, but this will require buy-in from academia, governments, and employers.

The open-source model that we enjoy in the software world today has helped reduce the cost of software and could be inspirational to the future model of education. As we digitize education, the cost of content delivery should approach zero. Specialized organizations could provide services to develop curriculums and credentials in the same way that companies provide packaging and support for open-source software. This is one idea that could substantially reduce the cost of higher education and help address the student debt crisis.

Most people think of a university as a place to get a higher education degree, but this is only one aspect of a much broader ecosystem. Recognizing that universities provide functions beyond content delivery and information exchange is essential. There is a social component that allows young people to experience independence and responsibility in a generally safe environment. The university may be a place where a student finds a spouse, and it is undoubtedly a great place to network and build friendships that last a lifetime. It also serves as a feeding system for athletics and is an excellent source of sports entertainment that will continue to garner our interest and consume our time. Many universities are research institutions, and research will retain its critical role in developing technologies.

We cannot count on the informational component of this ecosystem to be a primary source of revenue for universities in the future, as digitization will drive competitive pricing to zero. However, there are many other ways to monetize the university ecosystem, and the business model needs to change accordingly. For instance, athletics and sports entertainment can be monetized. The commercialization of research in partnership with the government and the private sector can also be monetized. Government subsidies will continue to play an important role. And universities with large endowments can fund their activities via investment resources. A combination of these and other yet-to-be-developed revenue sources will shape the future business model of universities.

An important matter to consider is the role universities will play in educating students for future jobs. As discussed previously, in addition to critical thinking and analytical skills, future jobs will increasingly require soft skills such as creativity, imagination, emotional and social intelligence, and empathy. Universities eager to prepare students to enter the workforce seem to have placed less emphasis on developing these skills in the last few decades. There is a crucial opportunity for improvement in this area.

We have also seen the importance of lateral thinking and how critical it will be in the future. Universities will have to do a better job of developing lateral thinkers instead of specialists. Artificial Intelligence will be able to do many specialized tasks in the future. Humans must collaborate effectively with machines, using intuition, creativity, and decision-making skills, and universities must play a role in improving these capabilities.

An education system aligned with our future needs requires innovative thinking to modularize and personalize learning. The model must be democratized, making a low-cost, lifelong educational pursuit available to the masses. We need to develop a modern—perhaps Blockchain-based—credentialing system to place students' credentials on secure, decentralized, distributed networks that can be accessed only by authorized organizations.

In the future, we will continue to see improvements in remote delivery and collaboration capabilities by adding Virtual Reality and Augmented Reality to existing systems. By leveraging AI, we can do a much better job matching market needs with curriculum development. AI will be able to assess students' skills and capabilities and help close developmental gaps in much more efficient ways. As we explore the frontiers of Brain-Computer Interfaces, the way we process, store, and access information will be nothing like what we see today, taking the entire concept of learning to unimaginable new dimensions.

In preparation for entering an environment where AI, automation, and robotics are everywhere, we need to rethink the role of education. Rather than imparting knowledge, education must transition to empowering students to identify problems and apply lateral thinking to find solutions. Skills such as adaptation, resilience, introspection, sociability, responsibility, ethical behavior, and empathy will become increasingly applicable in a volatile world that is constantly changing.

As we enter a world potentially marked by AI-produced abundance, we must develop a new socioeconomic model where all humankind benefits. When we get there, work may no longer be something we are obliged to do but something we choose to do for multiple reasons. In that context, the role of higher education may return to what was originally the purpose of a liberal arts education—to provide broad exposure to many subjects, developing well-rounded individuals who are exemplary citizens. This post-scarcity scenario is the subject of the next chapter.

## Takeaways

» The skills required to succeed in the future will change. Even though we don't know what the future jobs will be, we can look at the current signals and deduce that certain human capabilities, particularly soft skills such as empathy and lateral thinking, will be in high demand.

» Generative AI is the first massively adopted AI tool that is helping us become more familiar with and proficient in collaborating with AI.

» A broad, generalized educational approach where we feed our brain diverse experiences to create context and build empathy can help develop the uniquely human capabilities that will be beneficial in working side by side with AI.

» The current formal education model is outdated and needs to be restructured.

# CHAPTER 6:

# A Post-Scarcity Future

So far, we have seen that AI is advancing, and its proliferation will affect many different areas of our lives. We have considered how humans are uniquely different from intelligent machines and how we must take a different approach to education to obtain the skills necessary to succeed. Now, we turn the discussion to the benefits we can expect from an AI-driven operating model. We will also consider the struggles we will find in this journey.

The future is unpredictable. This chapter's purpose is not to try to predict the future but to look at current signals and trends to create a hypothesis about the future. Therefore, it is important to keep in mind that any trends discussed here are subject to inhibitors and accelerators that may alter their trajectories.

Most people would agree that AI will enormously impact the world at large. However, opinions vary significantly on what that impact will be. My hypothesis is that the future will look very different from the world we know today. I don't know how long it will take for this future to arrive, but I would argue that the journey has already started and that the changes ahead will come at a much faster pace than what we have experienced in the past.

My hypothesis is optimistic in its potential but realistic in the difficulties we will encounter to get there. It points to a future where intelligent machines could produce our basic material needs in abundance, resulting in a low cost of living for humanity, redefining the need to work, and allowing us to spend more time seeking higher purposes.

We will start our discussion by exploring how machine-created abundance may result in a low cost of living for humanity.

# Abundance in a Deflationary Environment

The challenge with a discussion about abundance and deflation is that these concepts contrast with our current reality. Many people around the globe still live in poverty, and inflation is a rising concern. We must acknowledge this dichotomy up front while keeping an open mind about the unprecedented forces of change we are experiencing today.

The ideas behind a future scenario where we enjoy abundance in a deflationary environment rest on the foundation of technology-driven exponential changes. The difficulty we face is that exponential growth is challenging to visualize and comprehend for our linear-thinking brains. The exponential curve looks flat in the early stages as if not much is happening, until it hits the inflection point. Today, we find ourselves at a point in the exponential curve where progress in reducing poverty and creating abundance is very gradual. For example, China has lifted over one hundred million people out of poverty, and Brazil has done the same for tens of millions in the last three decades. Food insecurity across the globe has gone from more than 20 percent to under 10 percent today, so we are making progress, albeit slowly. There is still much more work to be done before we can create abundance.

Given this background, it is understandable that there might be some skepticism about a discussion regarding abundance in a deflationary environment. However, I ask you to suspend any disbelief you may have. Hopefully, by the end of the chapter, you will be persuaded that there are more than just techno-utopian delusions behind the premises of a future marked by abundance and low cost of living.

Before we dive into these premises, first we need to provide some context to help guide the understanding of why there might be skepticism in the first place.

## Cognitive Biases

Let's start by discussing cognitive biases, defined as patterns of deviation in judgment that occur in particular situations. Humans are subject to dozens of cognitive biases influencing our decisions and beliefs. It is essential to recognize these biases and be aware of their potential influence on our ability to accept new ideas and form new opinions.

A few cognitive biases may lead to skepticism regarding abundance and deflation under current circumstances. The first is confirmation bias, a tendency to find and remember information that confirms our perceptions and to disregard information that opposes them. If you believe the world's current challenges are so profound that it is impossible to overcome them, you may find dissonance in this chapter that may feel uncomfortable.

Another cognitive bias to keep in mind is availability bias. This bias refers to the tendency to overestimate the likelihood of events with greater "availability" in memory, meaning we may be influenced by how recent or emotionally charged our memories are. We must remember that we have recently gone through considerable pain and emotional burden due to the pandemic, the war in Ukraine, inflation, and a potential recession. What is occupying our available memory is mostly negative feelings that run counter to the ideas of abundance and low cost of living.

Other cognitive biases may impact our perceptions. For example, anchoring is the tendency to rely too much on one piece of information when making decisions. The bandwagon effect is the tendency to believe things because others do. Negativity bias is our tendency to overestimate the likelihood of bad outcomes. Our need for control creates a mindset that tends to overestimate our individual capabilities but to underestimate the world's capabilities, causing us to become global pessimists. The immediacy leanings of our brains also result in a tendency to overestimate what we can accomplish in the short term and underestimate what we can achieve in the long run. The reason for pointing out these cognitive biases is to increase our awareness and help us recognize potential obstacles to envisioning the possibilities shared in this chapter.

# The Pessimism Engine: The Media-Activated Amygdala

The amygdala is a component of a group of brain structures called the limbic system, which plays a vital role in emotion and behavior. The amygdala is best known for its role in fear processing. When we are exposed to a fearful stimulus, information about that stimulus is immediately sent to the amygdala, which can then send signals to areas of the brain like the hypothalamus to trigger a "fight-or-flight" response. Research suggests that information about potentially frightening situations in the environment can reach the amygdala before we are even consciously aware of anything that could cause us to be afraid. Consequently, a fear reaction is initiated before we even have time to think about what is so frightening.[65]

Once stimulated, the amygdala becomes hypervigilant. The fight-or-flight response system increases our heart rate in anticipation of a need to act, dilates our pupils for better vision, and sends blood to our muscles for faster reaction. This system served us well as our early warning system in an era of immediacy when we had to react quickly to the threat of a saber-toothed tiger in the savannah. Much has changed since then, but the amygdala responds the same way to threats that have no immediacy, maintaining vigilance even when unnecessary. This survival mechanism may manifest itself in current times as pessimism.

In today's environment, with the unrelenting news media bombarding us with bad news to get our attention, pessimism sets in, and good news gets filtered out. Our natural tendency to focus on present-day threats creates a doomsday perception that elevates bad news disproportionately to the good news that surrounds us but that remains mostly invisible.

A recent survey by the World Economic Forum shows that 84.2 percent of people surveyed have a pessimistic outlook for the world.[66] This remarkable finding comes at an age when most people are living longer, wealthier, and healthier lives than a century ago. We have made enormous strides in every aspect of human well-being, from reduced poverty, malnourishment, and infant mortality to higher life expectancy and overall quality of life. We still have more work to do to extend these gains to every corner of the world, but the data does not support such a high level of pessimism.

In *Abundance: The Future Is Better Than You Think*, the authors interview Matt Ridley, a scientist who studies the origins and evolution of behavior. Our species' fondness for bad news caught his attention: "It's incredible . . . this moaning pessimism, this knee-jerk, things-are-going-downhill reaction from people living amid luxury and security that their ancestors would have died for. The tendency to see the emptiness of every glass is pervasive. It's almost as if people cling to bad news like a comfort blanket." Ridley contends that there is an evolutionary psychology component to this behavior. "We might be gloomy because gloomy people managed to avoid getting eaten by lions in the Pleistocene," he says.[67]

## The Stockdale Paradox

But there is more to the human psyche than a predisposition to pessimism. As stated before, we are complex and often ambiguous beings. The Stockdale Paradox was made famous in Jim Collins' bestseller *From Good to Great*. It captures our resilience in the most gruesome circumstances.

James Stockdale was a Vietnam War prisoner for seven-and-a-half years. Collins had read Stockdale's memoir and found its grim details hard to bear, despite knowing that Stockdale's later life was happy—he became a legendary soldier and statesman. Collins wondered, "If it feels depressing for me, how on earth did he survive when he was actually there and did not know the end of the story?"

Here is how Stockdale responded: "I never lost faith in the end of the story. I never doubted not only that I would get out but also that I would prevail in the end and turn the experience into the defining event of my life, which, in retrospect, I would not trade."

Collins asked him about the personal characteristics of prisoners who did not escape the camps. "The optimists," he replied. "Oh, they were the ones who said, 'We're going to be out by Christmas.' And Christmas would come, and Christmas would go. Then they'd say, 'We're going to be out by Easter.' And Easter would come, and Easter would go. And then Thanksgiving, and then it would be Christmas again. And they died of a broken heart . . . This is a very important lesson. You must never confuse faith that you will prevail in the end—which you can never afford to lose—with the discipline to confront the most brutal facts of your current reality, whatever they might be."[68]

Naïve optimism can be fatal, but so is the loss of faith. Stockdale's lesson is that we must keep faith while confronting reality. Confronting reality does not mean we must become gloomy pessimists. Faith, despite the circumstances, is a critical element to our resilience, survival, and growth. Before we get to abundance, we need to traverse a transition period. Transitions are often tricky and this one in particular will test our resolve. But if we can see what is on the other side and understand why the transition is so trying, perhaps we will gather the strength to confront reality while keeping the faith.

# The Linear Mindset

We have another cognitive challenge to overcome as we look into the future. Our brains are more comfortable with local and linear thinking. This mode of thinking served us well at a time when most everything we experienced was within a day's walk and changed very slowly. For many centuries, life was essentially the same from one generation to the next, and change just followed a linear progression.

But now, we live in an environment that is global and exponential, and we have difficulty adapting to this modus operandi. Globalization is easy to observe and understand. Today, any disruption in one corner of the world is communicated synchronously to another, creating ripple effects worldwide. Exponential is a bit more difficult to visualize because it happens gradually, then suddenly. We are fooled by the slow-moving progress of the early exponential progression, only to be surprised by its sudden explosion after it hits an inflection point.

Let's look at an example to illustrate the point. If you take a regular 8½" x 11" piece of paper and start folding it in half, you will double its thickness with each folding. You will be able to do this about six times and end up with a piece of paper about one inch thick—try it. Now imagine if you were able to fold the paper fifty times. How thick do you think the paper would get? Take a moment to think about this without looking at the answer, and write down your answer.

If you are like most people, you used a simple linear heuristic to estimate the thickness of the paper and used a factor of ten which would result in ten inches. You would be very wrong. A few of you may have perceived that doubling the thickness with each folding represents an exponential progression, but you probably still guessed a figure that would be less than 1,000 feet. You would also be wrong.

If you could fold the piece of paper fifty times, its thickness would be the equivalent of the distance from the earth to the sun! I know this is hard to believe, but it is a simple math exercise. It is hard to visualize because our brains struggle with exponential progression. We fail to see how quickly the doubling of the thickness grows once it passes the inflection point. The same difficulty impacts our ability to see the future and understand how exponential technologies will transform the world in ways we can't even imagine.

Our linear mindset prevents us from anticipating the speed and impact of change, so we fail to appreciate the range of possibilities. We extrapolate present conditions and trends as a way to predict the future, applying linear heuristics from simple mechanistic systems to complex ones that do not operate in the same manner. As Tony Seba and James Arbib observe in *Rethinking Humanity*, "Whether we are planning investments, education, social and environmental policy, or infrastructure spending, narrow linear mindsets blind us to the emerging possibilities and the pace and scale of change approaching—society is hurtling towards the future with a blindfold on."[69]

Now that we understand how our cognitive biases, pessimistic tendencies, and linear mindset impact our thinking, we can hopefully move on with our discussion while suspending any skepticism that prevents an appreciation for the possibilities ahead.

The research-based evidence gives us strong reasons to believe the future will be marked by abundance. However, the journey will not be easy. To get to an abundant future, we must traverse a transitional period marked by several challenges, such as natural disasters, economic difficulties, and human conflict.

Let's take the twentieth century as an example. We can see in retrospect that despite multiple setbacks, including the 1918 pandemic, a great depression, two world wars, and numerous natural disasters, this was a period of significant progress for humanity—a century in which infant mortality decreased by 90 percent and human longevity increased by 100 percent. During this time, food cost has dropped thirteenfold, energy cost twentyfold, and transportation cost a hundredfold.

The exponential growth and the convergence of technologies will generate lower costs and higher productivity, creating unprecedented abundance in a deflationary environment. However, Artificial Intelligence, robotics, and automation will continue to encroach on tasks that are performed by humans, creating unemployment. We will be able to produce more at a lower cost, potentially resulting in a new socioeconomic operating system that naturally makes a more equitable distribution of this technology-driven abundance. But during the transition, we will likely see additional inequality, so we must create short-term containment measures to avoid a societal collapse.

It is tempting to think that technology will create more jobs and that the complementing force of technology will more than outweigh job displacement. While it may be true that technology will create more demand as it produces abundance, the reality is that as machines become more capable in many areas of economic activity, humans will become increasingly less capable. The new tasks created from this additional demand will be performed by machines instead.

The resulting job displacement and unemployment is a cause for concern. It may very well create several challenges, especially during the transitional period. However, as we move beyond the transition, from a purely economic standpoint, we will benefit from a new production system that will create unprecedented abundance and lower cost of living, counterbalancing the negative impact of unemployment.

This production system is what Seba and Arbib refer to as a creation-based system as opposed to the extraction system that we have used until now to produce most of the goods consumed today: "This new production system is based on increasing returns and near-infinite supply . . . A creation-based system can produce near-infinite outputs once the infrastructure is built—limitless quantities of organic materials produced from the genetic information held in single cells and the large flows of energy produced from the sun, with just a few further inputs." In a creation-based system, everything we need is built from the ground up, working at the molecular level. The building blocks, such as molecules, photons, electrons, and genes, are widely available and can be combined through innovative processes to create new matter that can be used to our benefit at a very low cost. We can use AI to help us find inspirational patterns in nature to accelerate this creation process.

The technologies that enable the creation-based system are converging, producing extraordinary improvements in cost and capabilities across sectors, reinforcing one another, and creating disruption with ripple effects across society. For example, increased demand for and investment in electric vehicles are driving battery improvements, which in turn influence the electricity storage market, which then boosts the market for solar and wind energy, which creates a catalyst for additional battery improvements, further increasing electric vehicle competitiveness relative to fossil fuel-powered vehicles, creating a virtuous circle.

Advances in information processing and communications are at the heart of the exponential technology growth that brings change at velocity and scale beyond anything we have ever experienced. The cost and capabilities of many technologies, such as sensors, communication networks, computing, and robotics, are expected to improve by several orders of magnitude over the next decades.

These creation-based systems driven by exponential technologies can create abundance that would fundamentally change how we think about the economy and the meaning of work. Abundance does not mean a life of luxury. It means providing humanity with the possibility of escaping the need to work for sustenance so that we become free to pursue higher purposes. Abundance is not a solution to humanity's problems, as it will not change human nature. The human desire for power, social status, and political influence will continue to impact the way we relate to one another and behave. However, abundance can significantly impact socioeconomic conditions and, very importantly, affect how we think about work, time, and value.

In addition to abundance, we may benefit from technology's unstoppable deflationary force. The creation-based production system could drive the cost of our basic needs—energy, food, communications, transport, education, housing, and healthcare—toward zero.

Technology is redefining entire industries and business models, incentivizing them to digitize their operations and production processes. Anytime a business becomes digitized, it leverages the power of Moore's Law—the doubling of computing capacity every eighteen months—and its corresponding cost reduction. For example, the first whole human genome sequencing cost roughly $1 billion in 2003. By 2006, the cost had decreased to $300,000. In 2016, the cost reached $1,000, and today a Chinese company claims it can sequence the entire human genome for about $100. That is the power of exponential growth and its corresponding deflationary power.

In a deflationary environment, the general cost of living decreases, allowing more families to enjoy a higher standard of living despite lower wages and unemployment. Imagine the average American family living a middle-class lifestyle for $3,000 per year. This low cost of living seems like a utopian dream to most people, but that is because we have grown used to living in a scarcity-based inflationary environment, failing to understand the deflationary power of exponential technologies. As we saw with the human genome example, technology-driven deflationary forces are staggering. As technology continues to impact every aspect of our lives, we should expect the overall cost of living to decrease dramatically as long as governments let the deflationary forces run their course—this is an important caveat that must be taken into account.

According to Seba and Arbib, "The cost of the American Dream, thought of in terms of 1,000 miles/month of transport, 2,000 kWh/month of energy, complete nutrition (including 100 grams of protein, 250 grams of healthy carbs, 70 grams of fats, and micronutrients), 100 liters of clean water a day, continuing education, 500 sq. ft. of living space, and communications, could be less than \$250/month by 2030 and half that by 2035." I think Seba and Arbib's timing is overly optimistic—I believe it will be decades before we reach this level of cost reduction—but the point here is to be directionally correct rather than precise in timing.

It is easy to dismiss the premise that deflationary forces will reduce the cost of living when the headlines tell us otherwise. In January of 2022, *The Washington Post* announced that prices in December 2021 rose 7 percent compared to the year before, and inflation in 2021 was the highest in forty years.[70] It is important to remember that when we discuss the deflationary forces of technology, we are talking about long-term trends over several decades. Along the way, we will experience price volatility due to several factors beyond this discussion's scope. But suffice it to say that the current economic environment is quite unusual due to pandemic consequences and its ripple effects on supply chains and the economy at large. According to Nancy Lazar, Chief Economist at Cornerstone Macro, a firm specializing in macroeconomic trends, consumers used three years' worth of goods in eighteen months during the pandemic, driving up prices. As demand starts to normalize, prices should also begin to stabilize.[71]

Central banks react to short-term fluctuations to maintain stability, and governments use the tools at their disposal to achieve their short-term objectives. As we will see later in this chapter, in the long term, government intervention can significantly impact the economy's natural course. But for now, we would like to focus the discussion on making the case that the long-term trend is deflationary and that a higher standard of living based on lower cost is not only achievable, but highly probable.

# Family Budget Reductions

Let's take a look at a few areas of living expenses that consume a significant portion of the typical middle-class family budget today and how these expenses are expected to decrease over the next few decades.

### TRANSPORTATION

The convergence of two exponential platforms will significantly transform transportation and reduce costs: energy storage in the form of electric vehicles and AI in the form of autonomous vehicles. This transformation is happening faster than most people realize—it won't be long before you can hail an autonomous electric vehicle from your phone, just like you hail an internal-combustion-engine car driven by a person today via Uber, Lyft, or other apps.

This new mode of transportation is called "Transportation as a Service" or TaaS. It contains a fleet of electric, autonomous vehicles that can go 500 miles without charging and 100,000 miles a year with a small amount of maintenance. We will benefit from various vehicle classes that will be available at a moment's notice with the press of a button.

According to Whitney Tilson, an American technology investor and hedge fund manager, "It is estimated that TaaS will reduce the total costs of transportation by ten times compared to owning your vehicle. TaaS could drive down the costs of transportation to just 10 cents or less per mile." Today, AAA estimates that driving a traditional vehicle costs about 80 cents per mile.

TaaS will be available on a subscription basis for an estimated cost of about $150 per month for unlimited rides. For comparison, the cost of owning or leasing a vehicle, maintaining it, filling it with gas, and insuring it is, on average, five times as much or more. These savings do not include reducing space requirements for parking and storage and reducing costs associated with accidents, injuries, and lost human lives. It is estimated that TaaS will reduce accidents by 90 percent. In the US alone, the economic impact of car crashes is estimated to be $1 trillion, so accident cost avoidance in the US is estimated to reach $900 million.[72]

TaaS is not a futurist's dream. It is already operational today on a limited basis. Waymo, Google's autonomous vehicle division, is running in Phoenix, Arizona, where thousands of users benefit from this transformative innovation. Lyft has partnered with Aptiv, which has provided 100,000 autonomous rides in Las Vegas, serving more than 3,400 hotels, restaurants, and entertainment venues. Other cities like Boston, Pittsburgh, and San Jose are launching autonomous vehicle-hailing service pilots, confirming that this technology is a practical, functional new mode of transportation.[73] The challenge in achieving wider adoption is making autonomous vehicles reliable in more complex and unpredictable environments.

## HOUSING

Another exponential platform that will have a deflationary impact is 3D printing. This technology is disrupting the construction industry, producing housing that is more affordable and energy-efficient, reducing utility costs. In less than twenty-four hours, 3D printers can print the foundation and walls for a small home for as little as $10,000 today and potentially $4,000 in the future—a fraction of the typical construction cost.[74]

Several examples of this technology are already being used, creating affordable housing throughout the globe. California Mighty Building met the US building code standards to produce 3D printed single-family homes that cost a third of traditional houses. Using this technology, a Chinese construction firm erected a fifty-seven-story building in nineteen days.[75] Entire neighborhoods are being built with 3D printing in semirural parts of Latin America for families that earn less than $200 a month.[76]

Furthermore, as robotics capabilities advance, construction efficiencies will increase, and the cost of producing additional housing will drop significantly. For example, Balfour Beatty, a construction company, believes its construction sites will be human-free by 2050. The more construction automation advances, the less costly housing will become, turning affordable housing into a reality for many people within a few decades.

Additionally, intelligent systems that match supply and demand will generate the energy required to power our homes, reducing costs. Most energy will be generated by solar photovoltaic (PV) systems complemented by wind and other renewable energy sources. In many markets, the cost of PV is already below the marginal cost of fossil fuel electricity. Distributed energy combined with battery storage will allow localized production of energy that costs less than the transmission and distribution of centralized energy systems. This efficiency will enable the total energy cost to approach 1 cent per kWh, a 100 times reduction from current costs.

## FOOD

Forty percent of food in the US goes uneaten. The National Science Foundation (NSF) awarded American University a five-year, $15 million grant to address the issue. The principal investigator, Professor Sauleh Siddiqui, believes this waste can be reduced: "Instead of looking at the food system as a linear system where we plug gaps of waste in a straight line, we're seeking to transform it into a more circular system where we can reduce, reuse, and valorize all of the food that gets wasted," Siddiqui says.[77] Initiatives such as these are investigating innovative ways to remove inefficiencies from the food production and distribution processes, consequently reducing the cost of food.

Several emerging technologies are also contributing to increased efficiencies. Vertical farming, cellular agriculture, and biotechnology are a few examples. These technologies can produce food on demand at the point of consumption, eliminating one of the biggest challenges and the most significant contributor to the cost of food: distribution.

Vertical farming is the practice of producing food on vertically inclined surfaces that are stacked in layers in large structures like shipping containers or skyscrapers. Instead of soil, vertical farming uses aeroponic, aquaponic, or hydroponic mediums such as peat moss or coconut husks. Controlled Environment Agriculture (CEA) artificially controls temperature, light, humidity, and gases to produce food indoors, maximizing crops in a limited space in a sustainable manner. According to Allied Market Research, vertical farming will reach $24.11 billion by 2030, growing at 22.9 percent per year.[78]

Cellular agriculture produces animal-based products from cell cultures instead of directly from animals. The idea is to produce meat, milk, or other animal products without raising, slaughtering, and butchering livestock. Bioengineers Perumal Gandhi and Ryan Pandya, founders of the start-up Perfect Day, use a fermentation process to produce proteins from fungi inserted with a cow's genes, creating a liquid with properties similar to animal milk as well as ice creams and cheeses.

Precision fermentation could lead to the end of animal agriculture. According to the "Rethinking Food and Agriculture 2020-2030" report, "Nutritious food that initially replicates livestock proteins (milk and meat) will not just be an order-of-magnitude cheaper, but superior in every way—the food itself (taste, aroma, texture, mouthfeel, nutrition, and variety), predictability of quality, price, and supply, as well as the impact on health, animal welfare, and the environment. Food production will shift from an extraction model, where we grow plants and animals to break them down into the things we need, to a creation model, where foods are built up from precisely-designed molecules and cells."[79]

Using biotechnology, scientist Mark Post unveiled the world's first lab-grown beef burger in 2013 from tiny bundles of muscle fibers created from cultured cells taken from a cow. His company, Mosa Meat, can now create eighty thousand burgers from a sample of cells the size of a sesame seed.[80]

As these technologies continue to evolve, they will drive continued cost reductions in food production.

## HEALTHCARE

Technology advancements such as gene editing, which utilizes tools such as CRISPR, allow the change of a single letter in a string of DNA, potentially unveiling the cure for thousands of diseases at a fraction of the cost of traditional therapies. As mentioned earlier, the cost of sequencing the whole human genome has decreased from $1 billion in 2003 to about $100 today.

Nanotechnology will allow us to connect our vital signs from our bloodstream directly to our healthcare providers, creating much more effective detection and prevention mechanisms that dramatically reduce healthcare costs. By leveraging AI and telemedicine, doctors can scale their practices and reach many more patients, further reducing costs.

Stem cell research can potentially treat several chronic conditions such as heart disease, Alzheimer's, and diabetes, the cost of which threatens to cripple the entire healthcare system. In addition to reducing healthcare costs, new therapies allow people to return to normal activities much sooner, reducing the strain on caregivers.

Improvements in therapies, higher levels of prevention, and increased efficiencies in the overall system will reduce the significant healthcare cost burden that afflicts humanity today.

## EDUCATION

We saw in the previous chapter how the entire education system needs to be disrupted. We discussed how digitization could drive information delivery cost to near zero and how a model similar to open-source software could create a decentralized curriculum development and credentialing mechanism at a fraction of the cost of a college degree today. This change could significantly reduce education's cost burden on families.

# Opposing Forces

I hope I have convinced you by now that there are significant deflationary forces at play that, in the long run, should result in a dramatic cost-of-living reduction. I have provided examples of such reductions in five budgetary categories that account for more than half of a typical family's expenses. But it doesn't stop there. Converging exponential technologies will touch every aspect of our lives, creating abundance and driving deflationary forces.

However, it is essential to recall that we must cross the chasm before achieving the potential benefits of these exponential platforms. This journey is just beginning, as the platforms that will drive these changes have flourished recently. It will take a few decades before we see their full benefits, and in the meantime, we need to navigate through the transition.

This transitional period could be marked by volatility, chaos, and uncertainty. Government interventions to achieve stability could alter the natural course of economic forces. The problem is that governments don't always have the foresight to take actions that result in sustainable, long-term prosperity. Political forces drive short-term decisions that appease current distress by means that have negative long-term consequences—it is the proverbial "kicking the can down the road."

We see this in full display with the ballooning US debt. As of this writing, the US debt has reached an astronomical amount of $31.5 trillion, representing $246,867 for every taxpayer. In 1980, the US debt was the equivalent of 34.5 percent of GDP. Today it represents 120.4 percent.[81] Since the US dollar is the world's de facto reserve currency, this is not just a US problem but also a global threat.

The problem is exacerbated by the government's unceasing inability to restrain spending. In the fiscal year 2020, the US federal government spent $6.55 trillion, while tax receipts and other revenues trailed at $3.42 trillion. Consequently, the US federal budget deficit soared to a record $3.1 trillion.

How can the US government continue to spend more than it collects through tax revenues? The obvious answer is through debt accumulation, as described above. The not-so-obvious answer is a little old trick that governments have used for decades, the "printing of money." In the modern age, governments don't literally print money, but they use financial instruments that practically have the same effect. They inject liquidity into the economy, resulting in more money chasing fewer goods, leading to inflation.

Printing money is just another form of taxation that benefits politicians and the owners of capital while harming the average citizen. Politicians benefit because they don't need to be explicit about supporting higher taxation—they can continue with their vote-seeking narrative of lower taxes on the middle class—while supporting "hidden taxation" policies. The owners of capital benefit from the rising prices of assets. Regular citizens pay the price as the cost of living increases disproportionately to wages, worsening inequality.

We see a dichotomy of deflationary and inflationary forces at play. If governments didn't intervene, the decrease in the cost of living due to deflationary pressures would become apparent relatively sooner. However, current modern monetary policies favor an environment where inflation is maintained at around 2 percent through various financial instruments.

Only time will tell how these opposing forces will play out. But we know that deflationary forces are driven by exponential technology platforms that are unrelenting. We can't continue to fight deflationary forces by increasing debt and inequality indefinitely. We need a new socioeconomic model that will allow for sustainable growth that benefits humanity.

# Rethinking Capitalism

While technology promises to bring us unprecedented abundance, our current socioeconomic systems and the policies that have worked in the past will no longer serve us and take us to a prosperous future. We need a new way of thinking where we can all benefit from abundance, where poverty is eliminated, and where for the first time in history, humanity can take the time to seek higher purposes instead of toiling in meaningless labor to meet basic needs such as shelter, food, education, and healthcare.

Our mindset and belief systems have created the socioeconomic systems we know today. We come from a scarcity mindset where there are not enough goods to satisfy humanity's needs. This scarcity mindset created a belief system based on the principles of extraction, exploitation, and scarce resource hoarding. For centuries we have been concerned about accumulation, and our entire societal norms, culture, and behaviors incentivize accumulation and idolatrize those who have accumulated the most. In the classic words of Gordon Gekko in the film *Wall Street*, "Greed is good."

Recognizing the flaws of the greed-is-good way of thinking, particularly the inequalities and social ills created by a system that rewards accumulation, we have tried alternative models where a central government is responsible for an equal and fair distribution of all goods produced. Such systems have failed miserably due to government corruption, investment disincentives, and other unintended consequences.

One could argue that human selfishness will always lead to a desire to accumulate, creating an unsolvable conundrum where there will always be those who have too much and those who don't have enough. If this is true, no matter how much abundance we produce, humanity's basic needs will never be met entirely. But a compelling counterargument asserts that the desire to accumulate is a consequence of our scarcity mindset. We accumulate because our society admires such accumulation. We are fascinated by those with superior abilities to accumulate in a scarce environment.

But what happens when we can sustainably produce goods on demand for the first time in history without much effort? What happens when there is so much abundance that accumulation becomes an irrational, grotesque, repulsive behavior? In an abundant environment, anyone can accumulate. But the question is why. If society no longer thinks of accumulation as a demonstration of superior capabilities and accumulation is no longer admired but snubbed by society, will our selfish impulses compel us to continue accumulating, or will an abundance mindset completely change how we think and behave?

Another way to think about this is to look at the economics of the open-source movement. Why do people contribute to open-source? People are not motivated only by material goods or the currency that allows them to obtain such goods. In the open-source world, other currencies, such as one's reputation and the respect and admiration of peers and the community derived from contributions, are just as valuable. In a world where material goods are abundant, alternative currencies such as respect and admiration may become more valuable and drive human motivation.

Many scholars believe that taxation is the solution to a future without work, where there is the potential for significant inequality. Daniel Susskind dedicates an entire chapter in *A World without Work* to make this argument.[82] The problem with taxation is that it is an obligation. Obligations are undesirable and cause people to want to avoid them, sometimes using clever schemes and hiring taxation experts to search for loopholes. In other words, instead of incentivizing people to share, taxation gives a reason for people to avoid sharing. This contradiction works similarly to how rewarding people for playing turns play into work, turning an autotelic experience into an obligation and creating a nonintuitive disincentive. Taxation turns our desire to share, whether for personal satisfaction, societal admiration, or other more complex motives, into something to avoid. It results in people getting satisfaction from minimizing the obligation and therefore sharing less, having precisely the opposite effect of what it is intended to do.

In an abundant deflationary world where the cost of living approaches zero, abundance distribution will happen naturally without the need for taxation. This natural distribution works similarly to how we enjoy many benefits for free in the software world—many of the applications that we benefit from today are free or near free, while more sophisticated functions can be obtained for a premium price. The same concept could be applied to the general cost of living in an abundant world where costs drop exponentially due to the growth and convergence of technologies. Our basic needs could be met nearly free, but if we wanted a more sophisticated lifestyle, we would have to pay a premium price.

While taxation may not be a desirable long-term solution, it is crucial to recognize that we are not yet in a post-scarcity environment and still need to navigate a difficult transition. During such a transitional period, we may resort to taxation-based solutions such as the popular Universal Basic Income (UBI), government subsidies for basic needs such as education and healthcare, or other social safety nets. In the long run, these mechanisms may no longer be necessary as the cost of living approaches zero.

We are already seeing a movement toward reimagining some of the fundamental elements of our socioeconomic system. For example, people are starting to question the purpose of the corporate entities that drive our operating system today. Is the primary function of a corporation to increase shareholder value, as we have assumed for many years? Or should corporations elevate the needs of all stakeholders and make decisions that benefit all of humanity instead of a select few? The Business Roundtable, a group of chief executive officers of nearly two hundred major US corporations, has issued a statement defining the purpose of a corporation, challenging the notion that a corporation's function is to, first and foremost, serve its shareholders and maximize profit.[83] The concept of a B Corporation was created to ensure social and environmental performance, public transparency, and legal accountability to balance profit and purpose. These ideas are still evolving and may require refinement, but they show a movement toward creating a better socioeconomic model.

With an abundance mindset, humans will likely reach for a higher purpose, a topic we discuss in the last chapter. In this environment, the accumulation of goods and wealth may no longer be a reason for societal admiration. Instead, we may turn our praises to those who can express themselves in ways that inspire us by finding peace, joy, and fulfillment in helping humanity reach a higher level of conscience and civilization.

## The Jobless Society

If indeed we reach the promised abundance of exponential technologies, and all of our needs are met through a cost of living that approaches zero, an important question remains. What will we do with our time in a world without work? And what are the implications to society and our overall well-being?

In *Homo Deus*, Yuval Harari argues that technological advancement will lead to the rise of a useless society. To some, calling people useless may sound offensive. But what Harari is trying to point out is that some people may cease to be of economic value. They will not have any productive capabilities that can be compensated for in the marketplace, and they will not be able to upgrade their qualifications through additional education for several reasons, such as intellectual limitations, lack of motivation, or another general inability to keep up with the capabilities of the machines. Instead of useless, I prefer to refer to them as the jobless society.[84]

Suppose we achieve abundance, reduce the cost of living to negligible amounts through technology-driven deflationary pressures, and rethink capitalism so that the distribution of such abundance allows most human beings to afford their basic needs, such as shelter, food, healthcare, transportation, and education. Will humanity then achieve peace and happiness? There is much more to achieving peace and happiness than material needs, so I don't believe that would be the case. Furthermore, a job is more than just a source of income for most people. It provides identity, meaning, and purpose as well.

The question then becomes, Will technological unemployment deprive people not only of income but also of their significance? How will people find meaning when a significant source of it disappears? Many scholars believe that work is essential for human well-being. Alfred Marshall, an economic historian, proclaimed that "man rapidly degenerates unless he has some hard work to do, some difficulties to overcome." Sigmund Freud is credited with saying that human well-being depends on "love and work." Sociologist Max Weber believes that people attach meaning to work because of religion. He states that the Protestant work ethic comes from the idea that one must prove, through tireless and continuous work, that their soul is worth saving. This assertion is not a true creed of the Christian faith. However, Weber believes religion is a source of inspiration for work. He spoke of work as a "vocation," a "calling," a task "given by God."

Studies have confirmed the negative psychological and social impact of widespread unemployment. Marie Jahoda, a social psychologist, and her colleagues found that worklessness can result in growing apathy, loss of direction in life, and increasing ill will toward others. Work seems to provide a societal function that goes beyond its economic utility.

There are numerous accounts of people who work despite being very wealthy. Work seems to provide a sense of meaning, status, and social esteem that is more significant than wealth. Many retirees experience physical and mental decline shortly after leaving their work routines, reinforcing the premise that work has much more meaning than just the economic benefits it can produce.

Does joblessness necessarily imply a meaningless life? Are there other ways to obtain meaning, status, and social esteem that are not associated with work? The connection between work and purpose is not universal and seems to be a relatively recent phenomenon. In ancient times, work was considered degrading rather than meaningful. Aristotle wrote that "citizens must not lead the life of artisans or tradesmen, for such a life is ignoble and inimical to excellence." Further evidence that work may not necessarily lead to meaning and fulfillment is that only 36 percent of US workers and 20 percent of workers globally are engaged in their jobs.[85]

In his classic book *Man's Search for Meaning*, Viktor Frankl argues that people can find meaning not only in their work but also in simpler life experiences, such as appreciating nature's beauty, the arts, and other passive but enjoyable activities. He even argues that we can find meaning in suffering.[86] The political philosopher Hannah Arendt argues that we live in a "society of laborers which is about to be liberated from the fetters of labor, and this society does no longer know of those other higher and more meaningful activities for the sake of which this freedom would deserve to be won." As we find ourselves free from the need to work, we may seek meaning in other activities such as the arts, sciences, literature, and spirituality. Is it possible that our work identity has been a flawed construct that we can be liberated from?[87]

For most of human history, spirituality has played a critical role in our lives and our search for meaning. Our lack of knowledge and understanding has driven us to seek answers to what couldn't be explained with our intellect. Despite all the knowledge and understanding we have gained through several millennia, we still cannot find scientific answers to the most fundamental question that seems to be beyond our reach: the meaning of life. In a world without work, we will have much more time to ponder the more profound questions, taking us back to the same spiritual connection that has comforted us for a long time, helping us find meaning and purpose. In the last chapter, we will discuss reaching for a higher purpose. But before we dive into that discussion, let's look at the transition and adaptation ahead.

# Takeaways

» The unrelenting news media bombards us with bad news to get our attention, and we often miss the good news that remains mostly invisible. Our cognitive biases and linear mindsets can be obstacles to seeing positive changes, adding to the pessimism around us. We must deal with reality, but that does not equate to becoming pessimists.

» Current technical capabilities and their rate of improvement are sufficient to create a post-scarcity future where the cost of living decreases dramatically, but government intervention could alter this trajectory.

» Technology-driven deflationary forces could reduce the cost of most items of a middle-class family budget, including transportation, housing, food, healthcare, and education.

» A post-scarcity future could change how we think about capitalism and the purpose of work and corporations. It could also result in a jobless society that does not derive meaning from a job.

CHAPTER 7:

# The Transition

In the previous chapter, we considered the possibility of achieving a technology-driven, post-scarcity future that would change the course of history, allowing the average family to live comfortably at a low cost in an environment that changes the value of and motivation to work. This chapter considers the obstacles we currently face and offers suggestions to overcome them.

We are living in unprecedented times. We have never experienced change with such volume and velocity in the history of humanity. Emerging technologies are driving disruptions to business, society, government, and the economy, creating anxieties and apprehensions about the future. And this is only the beginning. As we have discussed extensively throughout this book, AI, automation, and robots are entrenching on tasks performed by humans at progressive and unrelenting rates. This entrenchment will drive technological unemployment and, before deflationary forces take effect, leave many wondering where their next paycheck will come from, how to pay their rent, how to put food on the table, and how to afford increasingly out-of-reach education and healthcare costs.

Today we are experiencing rising inequality, financial instability, social unrest, ecological degradation, cybersecurity breaches, data privacy concerns, and the interference of foreign states and radical ideologies threatening to undermine our representative democracy. On top of all that, a pandemic unlike any other we have seen in over a century has wreaked havoc, taking away our freedoms, destroying businesses, and impacting our fundamental human need to socialize with others. Understandably, people are overwhelmed, anxious, and afraid as the near-term prospects look discouraging.

The organizing systems based on the Industrial Revolution that allowed us to progress on so many fronts in the past are reaching their limits and will no longer carry us into a prosperous future. The industrial mindset of extraction and exploitation of human labor and Earth's resources is at its breaking point, putting us on the brink of social and economic collapse. The inflationary policies that artificially sustain asset prices by printing money and issuing debt cannot persist indefinitely.

We must get through this transitional phase before we achieve the promised abundance and low cost of living that technology has the potential to deliver. Crossing this chasm will not be easy, but knowing there is potential for a bright future will help us carry on. We must face the realities of where we are today to overcome the challenges ahead. Taking a brief look at history will help us understand past mistakes and hopefully aid in avoiding them in the future.

## How We Got Here

The transition we are living through seems to be reaching its most vulnerable point now as the consequences of our past decisions and behaviors are becoming more intense and pronounced. However, it may be helpful to think of this transition as an extended period that started as far back as the Industrial Revolution when modern life's first notions began to develop. Let's take a moment to examine this relatively brief portion of our history to understand and reflect upon how we got to where we are. History provides context that increases our understanding and has many lessons that can guide us as we move forward.

The last one hundred years have been remarkable. In the early 1900s, as pointed out earlier, we saw the flourishing of three key technology platforms that set the foundation for the modern world: electricity, telephony, and the internal combustion engine. This period also marked a significant transition from a dominant British Empire to a new position led by the emerging power of the United States. This transition evolved into the world known today.

Raoul Pal, a former hedge fund manager and Cofounder and CEO of Real Vision, has developed what he calls the Ultimate Macro Framework that helps contextualize the current macroeconomic environment and its future implications by looking back at historical events.[88] Pal's framework, summarized below, reminds us how a single trigger can unleash a series of occurrences that can profoundly impact the world.

By the turn of the twentieth century, the British Empire had started to fade. The British started engaging in a series of wars with Germany, which was becoming the new rising power in Europe. One incident, the assassination of Archduke Franz Ferdinand of Austria, heir presumptive to the Austro-Hungarian throne, by a nineteen-year-old Serbian nationalist allegedly triggered a series of events that led to World War I. Austria-Hungary was furious and, with Germany's support, declared war on Serbia. Within days, Germany declared war on Russia—Serbia's ally—and invaded France via Belgium, which then caused Britain to declare war on Germany.

World War I started with a single assassination and ended when the dominant powers created the Treaty of Versailles, the negotiated agreement that contained the peace terms. In that agreement was a war repatriation payment for Germany to account for the damage to Europe. Despite Germany's rising technological prowess, the massive repatriation amounts were beyond its means. Their solution was to debase the currency by printing money, which led to Germany's hyperinflation, collapse, and the rise of the Third Reich under Hitler's command.

Hitler's ambitions led to several invasions, and World War II ensued. The war created an enormous industrial machinery supply and the invention of multiple new technologies that would be leveraged for other peaceful purposes once the war ended. The end of World War II resulted in a euphoric period, particularly in the United States, which surged as the new world power.

What followed was a new era of prosperity propelled by the peaceful application of the industrial and technological capabilities developed during the war and the birth of the largest generation in history—the baby boomers. Seventy-eight million people were born in the USA, representing a 40 percent increase in twenty years, and the world at large did not stay too far behind, growing by 30 percent. This remarkable population growth, a new wave of fiscal stimulus, and the reconstruction of Europe and Japan drove the boom of the 1950s. The golden age of the 1950s and 1960s was marked by an increase in real wages, an improved standard of living, and a new appetite for consuming durable goods. America was the epicenter of this post-war prosperity.

We also saw the retooling of the global infrastructure and the creation of the rules-based world order system that included Bretton Woods, the monetary system that established the gold standard, and the US dollar as the global currency other currencies were pegged by. This period was also marked by the creation of the United Nations, the formation of the State of Israel, and the General Agreement on Tariffs and Trades (GATT), which minimized barriers to global trade. The intent was to create a new set of centralized rules, incentives, and global interdependency to avoid what had happened in the past.

From the late 1960s to mid-1970s, the baby boomers entered the workforce at the highest rate in history, with an explosion in consumption and a rise in commodity prices, particularly oil. The rising inflation led to the US abandoning the gold standard, resulting in the current fiat system. Since then, wages have stagnated in real terms, despite GDP growth.

The American Dream of the 1950s started to become unattainable for baby boomers. Asset prices began to rise, but wages didn't, especially for blue-collar workers. Families struggling to keep up resorted to dual-income arrangements as women entered the workforce massively. Given real-term wage stagnation and the increase in asset prices, even dual-income households could not achieve the elusive American Dream.

The only way for Americans to possibly access the good life, the standard of living their parents had, and the rich imagery portrayed by the media was through debt accumulation. Government agencies and Wall Street created market mechanisms that provided easy credit, and consumers became engulfed in mortgages, car loans, student loans, and credit card debt.

The end of pension systems and the creation of retirement plans that required investments in the stock market further exacerbated the problem as money flowed into equities, raising their prices. An easy stance on mortgage credit inflated real estate prices. Asset prices continued to rise as wages stayed stagnant, and debt accumulated further. The 1980s marked the beginning of the cult of the stock market and real estate leveraged by debt. Wall Street became the epicenter of the world economy.

Meanwhile, the Soviet Union collapsed, and China started to rise and open its borders. The original GATT became the World Trade Organization (WTO), pushing the idea of open free markets to the entire world. Global markets led to the offshoring of manufacturing jobs and, eventually, service jobs. Americans, rich by international standards, were now competing with laborers who were more than happy to perform the same work for a fraction of American wages. Additionally, automation took away many of the remaining manual labor jobs that paid good wages. Except for a lucky few who could migrate to professional roles, many in technology-related fields, most middle-class Americans found themselves in debt and low-wage service jobs. Many tried to fend for themselves by participating in the burgeoning gig economy, where there are no benefits and no guarantees of a stable income. The poor became poorer, the rich became richer, and the middle class started to shrink.

In 1987, in a desperate move to stabilize the stock market after the crash, US Federal Reserve Chairman Alan Greenspan lowered interest rates and injected liquidity into the economy, avoiding a recession. Interest rates became a tool in the central bankers' box used to manipulate economic cycles and reach their objectives of keeping inflation and unemployment balanced. On the surface all seemed well, but this has led to an ongoing debasing of the currency, which decreases the average worker's standard of living.

In 1998, overleveraged emerging markets started to unravel, and the Asian Tigers got decimated by stock market crashes. The problem is that their debt was in a foreign-denominated currency, US dollars. Unlike the US, these countries could not debase their currency to pay for foreign debt. This crisis spread to the developed world which, once again, cut interest rates to alleviate the problems.

Meanwhile, household debt as a percentage of income continued to expand in the developed world. The only way middle-class Americans could afford home ownership was through mortgages that required little money down. As home prices continued to rise, many Americans did all they could to buy homes before they became even more expensive and unaffordable. Wall Street created a financial mechanism where these mortgages were resold in the form of securities, and easy credit became the norm. It all unraveled in the 2008–2009 financial crisis when many people had their homes repossessed, and the entire financial system came close to an unfathomable collapse. This was the worst financial crisis since the Great Depression.

At this juncture interest rates were already low, so there was only so much central banks could cut. They had to very quickly come up with a new tool to save the system, which led to the advent of "Quantitative Easing (QE)," central bank's mechanism for printing money through the purchase of financial instruments such as treasuries and mortgage-backed securities. That was the only way to save a system that was overleveraged and about to collapse. Consequently, the currency was further debased, and the standard of living for middle-class citizens further eroded. QE drives long-term rates down, which forces financial entities to speculate on assets such as corporate bonds, stocks, and real estate in search of higher returns, driving up their prices and creating potential bubbles.[89]

While QE appears to be effective in preventing stock market crashes, it artificially inflates these assets. Wealthy people who own these assets benefit from price increases. The middle and lower classes, on the other hand, see their dream of home ownership slip away due to overinflated home prices, and saving for retirement becomes more difficult due to overvalued equity prices and low-yielding fixed-income instruments.[90]

With globalization, the idea of the American Dream, which includes home ownership, cars, appliances, healthcare, education, and vacations, quickly spread to all corners of the world and became the consumption standard everyone measured themselves by. Credit became a global addiction in search of the American Dream. The reality of today's indebted generation is that the American Dream is beyond reach for most people across the globe, including its inventors, middle-class Americans. This reality has resulted in an angry generation, forming division and populism on both ends of the political spectrum. Social media adds fuel to the fire, and Russia takes advantage of this new weapon to further divide Americans in an attempt to destroy our democracy. Putin and other dictators worldwide mock American democracy, pointing out our dysfunctions and inability to lead the world.

Next came the pandemic, a black swan that added to the already chaotic environment, leading to a stock market crash and a supply chain nightmare. The government realized the depth of the danger, given how overleveraged the entire financial system was. They did everything to avoid defaults so the whole system would not collapse. They went as far as giving money to households, even rich people who didn't need it, spending trillions of dollars to keep the system afloat, further debasing the currency and increasing inequality.

Modern Monetary Theory (MMT) posits that this vast national debt, currently at $31 trillion, will never get paid back and that this is OK as long as we can continue to service the debt interest. Supporters of MMT claim that since the government is the issuer of money, it can just create more.[91] Even expert economists cannot come to a consensus on MMT. Still, you don't need a PhD in economics to understand that too many dollars chasing fewer goods leads to inflation and lowers the standard of living for those who depend on stagnant wages. When governments create money out of thin air, that money must be paid back through taxation or inflation. Taxation is hard to implement due to political reasons, so inflation is the only possible answer. This is where we stand today—an artificially created inflationary environment. Central banks are desperately trying to undo the damage by increasing interest rates and potentially causing a recession. As of this writing, unintended consequences are starting to show in the form of bank failures.

# Societal Challenges

Our problems don't end at the social, economic, and political challenges described above. They run much deeper. David Brooks, the author of *The Second Mountain*, describes a series of crises that affect our society, starting with the telos crisis. A telos crisis develops when people don't know what their purpose is. "If you know what your purpose is, you can handle the setbacks. But when you don't know what your purpose is, any setback can lead to a total collapse," he writes. Stunningly, between 2012 and 2015, the number of young people suffering from severe depression increased from 5.9 to 8.2 percent. Brooks provided this critical insight: "When you take away a common moral order and tell everybody to find their own definition of the mystery of life, most people will come up empty."[92]

In a study for his book *The Path to Purpose*, William Damon found that only 20 percent of young adults have a fully realized sense of purpose.[93] Finding purpose is one of our highest callings and will become increasingly essential in an environment where the things that provide many people with purpose today, such as work, may be taken away. We will discuss the search for purpose in the next chapter.

Additionally, we are living through a family structural crisis. The traditional family structure that provided the foundational social support needed for healthy development has been shattered. We previously discussed the increasing number of families led by a single parent. Nearly 30 percent of households in America today are held by a single person, compared to less than 10 percent in 1950. Astonishingly, most children born to women under thirty grow up in a household lacking a father figure. The consequences are devastating. The suicide rate has increased by 30 percent since 1999. For young people between the ages of ten and seventeen in 2016, the suicide rate had increased by 70 percent in one decade. Every year, forty-five thousand Americans kill themselves. Additionally, seventy-two thousand Americans die every year due to opioids.

We seem to have lost our ability to trust, and this has created an alienated society. Brooks points out that less than 25 percent of Americans trust the government, compared to 75 percent in the 1950s. He also states that, according to the General Social Survey, only 32 percent of the general public and 18 percent of millennials believe their neighbors are trustworthy. This generalized distrust has created an ethos of "each person for himself," putting more pressure on self-reliance, accentuating loneliness that sometimes leads to a sense of despair.

Self-reliance and lack of trust have resulted in extreme individualism and tribalism. Tribalism is a way to connect people around a common foe, an attitude of "us versus them" that can lead to hatred. The scarcity mindset stimulates tribalism as life becomes a battle for scarce resources in a zero-sum game. Partisan identity fills the void left by traditional institutions such as family, neighborhoods, and religion. Once politics becomes a person's identity, the electoral contest becomes a struggle for existential survival, and extremism and fanaticism become acceptable norms. Hannah Arendt wrote in *The Origins of Totalitarianism* that people who have become political fanatics were experiencing loneliness and spiritual emptiness.[94]

Spiritual emptiness leads to a feeling of superfluousness, lack of hope, and inability to answer life's fundamental questions. Humans, unlike machines, have spiritual needs that require taking a chance to accept that there may be more to our existence than the merely physical—more on that in the next chapter.

So how does this all unwind? How do we get past the mess we have gotten ourselves into and cross the chasm to the promise of abundance? How do we get in touch with our purpose and reestablish community and trust?

# How We Move Forward

From a socioeconomic standpoint, what we have seen develop over the last one hundred plus years is a chain of events that has led to the over-leveraging of our financial system encompassing a debt level that has ballooned to unprecedented heights. I believe there are only two ways to resolve this problem. One is to follow the Austrian economic school of thought, basically the nuke option. Press the reset button and let the cards fall where they may. It is the equivalent of declaring global bankruptcy. The pain this would create is unconscionable.

The other option is to reverse the entire process. Let's review how this would work. We are living on borrowed money. More dollars are chasing fewer goods, eroding purchasing power and lowering the standard of living for those not lucky enough to benefit from increased asset prices. So how do we reverse it? If you are a household going through a debt problem, you know that you only have two choices: consume less or produce more. Consuming less is painful, and producing more is difficult and takes time. The saving grace is that inflation can be countered by deflationary forces powered by technology-induced productivity, so long as we can buy enough time for these deflationary forces to work. The danger in an overleveraged environment like the one we live in today is that when the dominoes start falling, the entire system could collapse, halting the process that would have led to the solution and causing considerable pain.

In the long run, we will be in a position where the exponential growth of technologies can sustainably produce more than we consume. This deflationary force will counter the inflationary pressures created by central bank policies. It is hard to predict how this colossal battle between inflationary and deflationary forces will play out. Still, we know that exponential technologies are unrelenting, so in the long term, they likely win.

The problem—and here is where we need to focus—is that we have a chasm to cross before we can reach the abundant side of history. If we cross the chasm successfully, we will be in a position to forgo our economic worries and focus on more important matters. So, what do we need to do to cross the chasm successfully? We need to apply change management principles that have been deployed successfully in the business world to help guide us through this process.

Here are specific action items that we must take to cross the chasm:

## SET A VISION

We are going through massive, fast, hard-to-absorb changes. By studying change management literature, we learn that successful change starts with a well-articulated vision that motivates at least the portion of the target audience who believes in it. In any group of people, you will always have detractors trying to interrupt change and skeptics taking a wait-and-see attitude. But you also find champions, people who believe in the vision, are willing to lead, and can bring others along.

We need to set an optimistic vision for our future. The pendulum has swung too far to the pessimistic side and is entirely out of balance. As discussed previously, we don't need to become Pollyannas—we must face reality as it is. However, we must not lose faith. We need hope, courage, and resilience. It starts with a vision for a better future, which I sincerely hope this book has contributed to.

## DEVELOP LEADERSHIP

There is so much negativity today and so much division. Much of it is driven by fear and lack of leadership. We need a new breed of leaders who can paint a picture of a prosperous future and articulate a credible plan to get there. The new generation, which is much more in tune with technology and more conscientious about the disastrous consequences of our past decisions, may provide the seeds of leadership we need. We must develop capable leaders and vote for people of character, courage, and wisdom who are fit for the responsibilities of the offices they hold.

## PROVIDE EDUCATION

We need more education. We need to help people understand that there is a prosperous future if we use the tools already accessible to us right now. We don't need any significant technological breakthroughs—the technologies that we have today and their rate of improvement are sufficient to get us to a brighter future.

The problem is that most people are unaware of this, and most people don't understand exponential growth. All they see is scarcity, inequality, and environmental degradation. This blind spot creates much fear and division. We need to educate people so they can understand the possibilities ahead and work together toward a thriving future.

## THINK DIFFERENTLY

To make progress, we need to embrace a new way of thinking. We must recognize the complexity, magnitude, and velocity of the changes ahead and adopt a systems-thinking approach to problems, avoiding the temptation of applying simplistic point solutions without considering unintended consequences.

We must embrace change and let go of our fears. We must adopt an abundant exponential mindset. We must let go of old orthodoxies and legacy institutions that no longer serve our needs. We must rethink our entire education system. We must embrace technology-driven changes that are unstoppable. Organizations must adopt innovative methodologies and abandon the old ways of doing business. Individuals must embrace change, find community, and become self-sufficient through entrepreneurship.

## STOP THE BLEEDING

We must stop the bleeding. We must stop the destruction of the only home we have, Mother Earth. We must maintain the economic system functioning but avoid the temptation of printing money ad infinitum leading to the collapse of the entire financial system.

We need a better monetary system that replaces outdated central bank mandates to keep low unemployment and 2 percent inflation. We know that the unrelenting forces of exponential platforms will drive unemployment and deflation, and resistance is futile. Fighting technology-driven unemployment and deflation with fiscal policies is like trying to stop a tsunami with sandbags. We have already seen the unintended consequences of tools like Quantitative Easing resulting in the surging inequalities we are experiencing today.

## CREATE SAFETY NETS

We need to establish social safety nets for those who will be hurt the most during this transition. We must maintain the inviolability of our social security and healthcare promises. It is essential to support social, economic, and political stability to avoid collapse.

Much has been discussed about Universal Basic Income (UBI). There are many challenges in developing such a system. The temptation is that UBI will be funded by printing additional money, which leads to more inflation. The best way to distribute wealth is through deflation. If the cost of living decreases, everybody benefits, including those who depend on stagnant wages or resort to temporary gigs to stay afloat. But until we reach the point where deflationary forces are strong enough to reduce the cost of living significantly, we may have to resort to inefficient mechanisms such as UBI as a temporary stopgap.

## BUILD RESILIENCE AND SELF-SUFFICIENCY

We need to continue developing capabilities that are robust and resilient. Recent events have demonstrated the problems associated with high dependency on global supply chains. The disruptions associated with COVID-19 were a wake-up call that alerted us that we were going in the wrong direction. For essential needs such as food and energy, we must develop local self-sufficiency and decentralized distribution systems that are not subject to single points of failure.

We must leverage technology to create new opportunities and new income sources to replace employment-based wage dependency. We need to encourage our youth to develop technical competencies and business knowledge so they can take advantage of the ongoing demand for innovative entrepreneurial solutions.

In essence, we need to continue moving toward flexible, resilient, transparent, open-source systems where everybody benefits, not just the privileged few.

## STAY UNITED

We must stay united. As Abraham Lincoln eloquently stated, "A house divided will not stand." We need to go back to a civilized discourse, a healthy debate. We must not let traditional and modern media fueled by foreign interference divide us and destroy the democratic freedoms we have fought so hard for. We must realize that the enemy is not the opposing party. The enemy is those who want to dismantle our democracy. The enemy is the irreversible destruction of our planet. The enemy is our inability to open our minds and have civilized debates with those who think differently.

## OVERCOME FEAR

Fear is at the root of much of the division we see today. Change is uncomfortable, but once we understand and believe that what awaits us on the other side is good, there is nothing to fear. As Franklin Roosevelt put it: "The only thing to fear is fear itself."

# We Have Choices

We find ourselves in a period of turbulence, instability, and transition. We must be resolute in our willingness to get past this transitory period by relinquishing policies, attitudes, and orthodoxies from our industrial-based exploitative mindset. Then, when we get to the other side, we will find abundance unlike any we have ever experienced. We must find the pathway to cross the chasm.

As Tony Seba and James Arbib eloquently put it in *Rethinking Humanity*, "For the first time in history, we have not just the technological tools to make an incredible leap in societal capabilities, but the understanding and foresight to see what is coming. We have the choice, therefore, to avert disaster or not."

Successfully crossing the chasm requires understanding the challenges and opportunities ahead, applying change management knowledge to help us adapt, and embracing a new abundance mindset. The extractive production systems resulting from the scarcity mindset have been ingrained in our society for two hundred years. But these systems are incompatible with the creation-based model. Using the same old methods to navigate a new paradigm will not produce the desired outcomes and may create additional political divisions, inequality, and social instability. The danger is that our inability to understand the complexity and depth of the changes we are about to experience may push people toward simplistic, extremist, populist solutions that are ineffective.

Trying to fix the current systems with temporary patches and looking for solutions rooted in a linear mindset will only worsen the problems. Solving climate change by limiting consumption has negative social impacts. Solving food insecurity by cutting trees to create additional farming land exacerbates the environmental issues. Solving inequality with taxation creates investment disincentives with negative economic consequences. We must let go of the old ways of thinking.

Throughout history, every civilization that reached a breaking point like the one we are facing today has collapsed. But we don't have to repeat the mistakes of previous civilizations. We have more capabilities today than we have ever had in the past. Today we can harness AI to help us make better decisions by running numerous simulations to estimate a range of possible outcomes as we consider alternatives that produce the best results. The ability to make informed decisions based on simulated experiments is a game-changer. And it is within reach with the technologies we have access to today.

Crossing the chasm from the current state of chaos and instability to the other side, where we find the promise of abundance, lack of inequality, and the ability of the human spirit to rise above the mundane, will take all of our collective efforts. We must unite as a species, overcoming the division caused by intransigent political orthodoxies. We need to eliminate racial injustice, rethink the current economic operating system, stop destroying the environment while there is still time, and make the difficult choice to stop kicking the can down the road. We must consciously accept the short-term pain of reducing global debt before it gets out of control. We must be resolute in confronting our current realities without losing faith to get to the other side without a total socioeconomic collapse.

The question is, Can we, and will we? In a quest to answer this question, we find hope in social psychology experiments done from the 1950s through the early 1960s. In his book *The Robbers Cave Experiment: Group Conflict and Cooperation*, Muzafer Sherif describes an experiment to study intergroup conflict and resolution between groups of eleven- and twelve-year-old boys. In the experiment's initial phases, Sherif assigned kids to groups and established activities that created group identity. He then introduced competition between groups and observed that the competition created conflict and hostility. The boys developed "us versus them" group identities and turned into hostile, narrow-minded opponents even if they didn't have any previous behavioral issues. This animosity may not be surprising. But what is interesting in Sherif's observations is what happened in the last phase of the experiment, when he introduced problems that impacted all the groups and that required collaboration among them to solve—he called them superordinate goals. These superordinate goals resulted in the groups becoming closer to each other and eventually forming close relationships.[95]

We may be able to cross the chasm successfully if we find superordinate goals—such as addressing environmental degradation, pandemics, and other common afflictions—to help us overcome our current divisiveness. An increase in the frequency or magnitude of such events may be the spark that ignites us to unite around these superordinate goals. Getting through the transition will require leadership and resolve. But if we do, we will have the opportunity to experience a new paradigm of human existence focused on the higher purposes sought by our souls— that is the focus of the next chapter.

## Takeaways

» Our reality includes financial instability, rising inequality, social unrest, ecological degradation, and many other ailments.

» History provides context that helps us understand how we got to this point, which can be useful in developing effective solutions and avoiding past mistakes.

» The change management discipline can guide us as we cross the chasm from the current environment to a brighter future.

» To successfully arrive at the other side, we must set a vision, develop leadership, provide education, think differently, stop the bleeding, create safety nets, build resilience and self-sufficiency, stay united, and overcome fear. We have choices, but may need to find superordinate goals to overcome our divisions and make the right decisions.

# CHAPTER 8:

# Reaching for a Higher Purpose

As previously discussed, there is something unique about being human that differentiates us from machines. Machines may one day be able to become superintelligent and surpass us in every intellectual capability. Artificial Emotional Intelligence may be able to mimic human emotions. Automation may take away the jobs that keep us busy, provide income, and offer some people an identity. But what the machines can't take away is our free will and innate need to find meaning and purpose—machines cannot duplicate our souls.

Humans are not just flesh-and-bone automatons programmed by DNA, driven by neurochemicals, and directed by neuron synapses firing in reaction to momentary circumstances. We are conscious. We need to know why we are here, where we came from, and where we are going. There is a more significant meaning to life that is beyond our understanding. We have a soul that seeks a relationship with others, the universe, and our creator.

In chapter 4, I introduced the work of Drs. Iain McGilchrist and Sharon Dirckx in our discussion about the relationship between consciousness and the brain. I also pointed out that philosophers have pondered the question of who we are for centuries. This is a fundamental matter we need to come to grips with to understand our true identity and what distinguishes us from intelligent machines. It also has critical implications for what it means to reach for higher purposes. This discussion brings us back to Dr. Dirckx's book *Am I Just My Brain?*, which lays out an excellent framework for deriving a sensible answer to this fundamental question. My goal here is not to provide an extensive argument defending why I believe we are more than just our brains. For that, I highly suggest you read Dr. Dirckx's book. But to give context to the discussion about humans having souls that cannot be imparted to machines, it is helpful to extract some key ideas from Dr. Dirckx's beautifully articulated arguments.[96]

We can begin by revisiting the fundamental issue of identity. Who are we? Advanced primates? Machines? Our brains? Do we truly have free will if electric currents and chemical reactions in our brains drive us? If we build intelligent machines also driven by electric pulses, are they also conscious? If we are just our brains, do we gradually lose our sense of self as our brains decline with age? Worldviews vary on the identity of a person. The fashion industry says we are our body, the financial sector says we are our income and wealth, and neuroscientists say we are our brains. But industry and science cannot answer the question of identity alone. It requires the perspectives of philosophy and theology.

Considering this broader perspective, it is essential to distinguish between the mind and the brain. The mind is a broader concept that encompasses a person's inner life, thoughts, feelings, emotions, and memories. Mind and brain are related, but exactly how is the question that has been occupying philosophers, ethicists, and theologians for centuries. One view favored by modern scientists is called "reductive physicalism," which states that the mind is reducible to the physical workings of the brain. An alternative view called "non-reductive physicalism" says that when parts of the brain combine and reach a certain level of complexity, they give rise to something new and distinct that cannot be reduced back to its original components. Another view, called "substance dualism," states that the mind is beyond the brain. They are two separate substances that interact but can also operate independently.

What is this other substance beyond the brain? Science can't answer that question because it limits its perspectives to what can be seen and verified. The problem is that we are of the same composition as the subject of our study and, therefore, can't conform to the objectivity requirements of science. This dilemma gets us nowhere. However, if we are open to the possibility that there is more to the mind than we can see, perhaps we can find answers.

If we are more than just our brains, what more is there? We have commonly referred to this additional part as the soul. But what exactly is the soul? We use this word casually to refer to things like soul music, soul mates, and soul-searching. The term soulful can even be interpreted as a sorrowful feeling, which is not the intended meaning of this book's title. However, beliefs that the soul is the essence of a person have been around for centuries. They have been vital to promoting equality, ending slavery, and defining other important ethical concepts that have been made into law.

The idea of the soul goes back to ancient Greece, where the philosopher Plato argued that the soul is the ultimate source of life for all living things, and for humans, the soul is the essential person who lives on even after the body dies. Aristotle studied under Plato and believed that the body and soul are integrated and that a particular soul occupies a specific body. Another Greek philosopher, Pythagoras, had a different point of view, believing in reincarnation and the possibility that the soul could occupy other bodies, influencing Eastern beliefs held by Hindus, Buddhists, and Sikhs. The Bible, which guides the Christian faith, speaks often about the soul. According to the Bible, the soul is given by God to each of us and is the essential core of who we are. Jesus taught us that we can enjoy eternal life and that our souls are indestructible in this world.

There are voices within neuroscience and philosophy today who believe there is no soul and that we live in an entirely material world.

They ignore the conclusions of great minds that have thought deeply about this issue over millennia. However, many philosophers and theologians today consider the soul to be the immaterial core of a person that integrates everything else: the mind, the will, and the body.

In 1950, Alan Turing devised a methodology to measure the intelligence of a robot known as the Turing Test, which would help us distinguish a person from a machine. Today robots can not only pass the Turing Test but also convince humans that they are conscious, as was the case with the Google engineer who declared that his AI was sentient. *The Terminator* series starring Arnold Schwarzenegger depicted conscious robots that would eventually overcome and obliterate humanity. Given this context, we can't help but ask: Is science fiction a predictor of the future? Will robots eventually take on a mind of their own and become indistinguishable from humans? Will they gain the equivalent of human consciousness and demote humanity to an inferior status in the universe?

Dr. McGilchrist tells us the story of the master and his emissary that inspired the title of one of his books. It comes from a fable of a wise spiritual master who looks after the community which thrives because of his wisdom. But he also knows that as the community grows, there are details he shouldn't get involved with so that he can provide oversight and maintain his pristine role as master. Therefore, he appoints a second-in-command, the emissary, who carries out his procedures as an efficient bureaucrat. The emissary thinks he knows everything but doesn't know what he doesn't know. He then starts acting as if he is the master and chooses not to get the master's council and guidance on matters he is not knowledgeable about. Consequently, the situation deteriorates and falls apart. Dr. McGilchrist explains that versions of this story exist in many cultures around the globe, including China, Japan, and India.[97]

The moral of the story is the danger of the emissary becoming too powerful and not accepting his subordinate position as a faithful servant, not realizing what he doesn't know. As we reflect on this story as it concerns AI development, a couple of questions surface: First, have we stepped out of our boundaries in creating an intelligent entity that could potentially overcome humans? In other words, are we playing God, and given that we don't know what we don't know, will that be humanity's downfall? Second, will AI someday step out of its boundaries as a faithful servant of humanity and start taking on the role of the master?

There are no definitive answers, but there are ways to think about these questions that might provide guidance. If you believe, as many of today's neuroscientists do, that we are nothing more than what happens in our brains, then I think it would be easy to come to the next logical conclusion that intelligent machines will someday overcome us, dominate the world, and potentially destroy us or put us in a subordinate position. However, if you are open to the alternative view that we are more than just material beings and have a soul given to us by our creator, there is hope. Is it possible that we have opened Pandora's box, and that intelligent machines will lead to our earthly destruction? Maybe. But even then, if you believe in a merciful and loving God who is much more expansive than our limited humanity allows us to comprehend, there is hope even beyond the natural world's realm.

I like to take an optimistic view that human-AI collaboration will lead to a prosperous future, as evidenced in arguments made throughout this book, despite this being an unpopular view. According to a Monmouth University poll, more than half of Americans are very or somewhat worried that AI could one day pose a risk to the human race, and only 9 percent believe it will do more good than harm. I humbly admit that I could be wrong about my optimistic view. It is also possible that we will fail to get through the transition. Regardless of what happens, faith in God's promises of eternal life can bring us peace.

Suppose we cross the chasm and arrive at the other side, finding ourselves in an environment where machines provide most of our needs. In that case, we may have an opportunity to gain a new appreciation for our time and values as we seek to reach for higher life purposes. Next, I propose several transformations I believe society may experience as we journey into this post-scarcity environment devised by intelligent machines.

## Employment to Expressive Works

As discussed throughout this book, technologies will continuously encroach on tasks currently performed by humans. Most of what we think of as work today will likely be performed by machines and become commoditized.

Economists used to think that value was fixed, like the production output from agriculture and manufacturing. But economists like Mariana Mazzucato are challenging that notion, and more people are questioning preconceived ideas of value. As automation commoditizes work and production, value may become a more fluid concept that accrues to human activities that reflect creativity and originality.

Entertainment will likely continue to be an essential part of our lives if we are freed from the toil of labor due to automation and creation-based production systems. Adam Smith thought that opera singers, actors, dancers, and the like were frivolous and created no value for society. However, many of the most highly paid professionals today are entertainers: actors, musicians, athletes, and social media influencers. Creativity has moved to the heart of today's internet-fueled "attention economy."[98] This movement toward human expressive works becoming more valuable relative to commoditized production will likely become more prevalent as machines encroach on all forms of work.

A unique characteristic of expressive works is that they take time. We can't rush or schedule inspiration. This is in drastic contrast to the way we think about time today.

## High Time Preference to Low Time Preference

In economics, time preference is the concept that people prefer goods available for use at the present over goods becoming available at some time in the future. For example, future cash flows are reduced by a discount rate to arrive at the net present value because cash now is more valuable than cash in the future.

The concept of time preference becomes apparent in a hyperinflationary environment, as some countries have experienced. If the price of goods tomorrow will be much higher than the price of goods today, you are incentivized to spend your money today. This shortens the time span one operates in. In this scenario, people don't worry about saving for the future because those savings become less valuable than the benefits they provide today. In a hyperinflationary environment, people have a very high time preference.

We live in a world of short attention spans, busyness, and the desire for immediate gratification. We have accumulated enormous debt to obtain what we want now, even if we can't afford it. Dual-income families have become the norm because it takes two full-time earners to keep up with modern life's demands and meet our cravings for immediate gratification. We live in an environment that incites high time preference driven by our materialistic scarcity mindset.

The migration to an abundance mindset and from employment to expressive works stimulates low time preference. Deflation is conducive to low time preference as it incentivizes savings and delayed gratification. In a deflationary environment, tomorrow's dollars are more valuable than today's. Busywork will be performed by machines and become commoditized, affording us the luxury of doing inspirational work that requires time. Increased life spans and healthier aging brought by advances in biotech and medicine can also contribute to this shift to a low time preference as we gain more time in the future.

# Hedonism to Eudaimonism

Hedonism is the practice of seeking pleasure as a source of happiness. It is also manifested in the form of immediate gratification. A high-time-preference lifestyle induces hedonism because when our time span is short, we try to maximize happiness by engaging in pleasurable activities, even though we know this happiness is not long-lasting. Hedonism is a form of escapism, a way to focus on present satisfaction and gratification while disregarding future consequences.

On the other hand, eudaimonism is the practice of living a life that leads to long-term joy and contentment rather than short-term pleasures. A low time preference is conducive to eudaimonism since it provides a long-term perspective in which we maximize long-lasting joy.

## Extrinsic Rewards to Purposeful Pursuits

Earlier in this book, we learned about the Sawyer Effect. If you provide an extrinsic reward to someone to perform a task that the person would naturally consider play, you may turn the play into work and create the opposite effect of what the reward intended to achieve. We have also discussed autotelic experiences in which the experience itself is its own reward. We have learned that the highest, most satisfying experiences in people's lives are autotelic, putting them in a state of flow. This discovery is critical in understanding our motivation to perform work in the future.

In algorithmic work—the type of activity where you follow established instructions—extrinsic rewards work well. But this work will primarily be performed by machines and will become commoditized. The activities that will remain available for humans to perform are the types of tasks that require inspiration, artistic expression, and originality, all of which require time.

Purpose and intrinsic motivation will carry a progressively higher weight as we consider how we occupy our time. Increasingly, people will be more purposeful as they consider the jobs they want to do and the companies they want to work for. As work becomes automated and commoditized, our motivation will be driven by deeper reasons, leading to the pursuit of activities that are autotelic or that fulfill a higher purpose.

## Materialism to Spirituality

In modern times we tend to put a tremendous amount of emphasis on our material needs. In Western cultures, there is an obsessive desire to find fulfillment through success, fame, money, power, prestige, and other worldly matters. We seem to have lost our ability to believe in the unseen, limiting our belief system to what science can explain. However, as people approach midlife or encounter times of human calamities such as wars or pandemics, we seem to become more open to accepting that there is a spiritual dimension to our existence.

The evidence is clear that people find happiness and greater life satisfaction in faith and spirituality. In an episode called "Searching for Spirituality," in the podcast *The Art of Happiness with Arthur Brooks*, Brooks suggests that many people who have not had any religious leanings in their lives or do not profess to have faith have a certain curiosity about religion but are embarrassed to admit it. They may approach someone with some knowledge about faith in a subtle way, almost sheepishly. Why would people be embarrassed to ask questions related to faith or religion?

Brooks did some research to learn more about this curious condition and was shocked by his findings: "It turns out that people have a tendency in their thirties, forties, and even fifties and beyond to have religious inklings for the first time, to have questions, to have urges, to find that they have interests and maybe even needs to go deeper into their own spiritual life, sometimes for the very first time in their lives. And don't know how to deal with it."[99]

James Fowler, an academic, theologian, and social scientist, wrote a famous book called *Stages of Faith*. In the book, Fowler describes that he discovered that people in their middle age start to have religious urges for the first time. This experience can be extremely disconcerting to people who, for their entire lives, were apathetic or even opposed to any religious inclinations. Fowler's research found that, for the most part, children don't find religion very interesting until they hit the age of twelve. Between the ages of twelve and eighteen, they tend to conform to the religious views of their family. As they reach young adulthood, they go through what Fowler calls the Individuative-Reflective stage, characterized by angst and doubt. This is a time when these young adults start to become more independent in their thinking and beliefs. They struggle with believing in a loving and merciful God in a world full of pain and suffering.

In the Conjunctive phase, which tends to start in the midthirties, people begin to resolve contradictions better. They realize that the beauty and comfort of growing in faith outweigh the inconsistencies. They become more prone to accepting their lack of understanding and stop trying to explain the unexplainable. They embrace the mystery as just part of our existence as humans and humble themselves in light of the majestic universe that we live in.

A conflict arises when people who have made an intellectual or philosophical commitment to rejecting faith start to seek a deeper meaning in their lives and desire the beauty and majesty of faith despite any inconsistencies that it may represent. However, at this stage in life, people can live with this dissonance that they were uncomfortable living with before, and the inconsistencies are not of as much consequence as they once were. The conflict between the intellect's rigidness and the soul's desire causes the embarrassment described earlier. The final stage of the faith journey is what Fowler calls Universalizing when people later in life become enlightened and comfortable with their faith.[100]

Surprisingly, during periods of war or pandemics, we find that people suffer less from the "diseases of despair," such as depression and substance abuse. The explanation is that during these difficult times, people tend to focus less on themselves and more on things of spiritual nature that fulfill the needs of the soul.

Research suggests that faith leads to a happier life. This can be explained by the fact that pondering about the metaphysical or the transcendent removes the focus on the self. Major depressive disorder is associated with an obsessive focus on a person's own problems. Switching one's focus from the self to the transcendent or metaphysical is in and of itself therapeutic. Harold G. Koenig, a psychiatrist on the Duke University faculty, has done an exhaustive review of religion's benefits, and he has found that religion and depression are negatively correlated. However, to benefit one must engage and not just declare adherence to any particular faith or belief.

Why do we have this natural tendency to seek and understand religion, especially as we mature and gain life experience? Can science explain religious tendencies? The Cognitive Science of Religion (CSR) studies what is happening in the brain during religious belief and practice, and it is a growing research area. According to CSR, humans have a natural bias toward religious beliefs, and it tries to explain why. It argues that in the Dark Ages, religion was seen as mysterious and supernatural, but now we can show that it is a natural phenomenon that happens entirely within our brains. It tries to explain away religion in three possible ways: human error, a product of evolution, and genetic predisposition. The argument suggests that religious beliefs provide cohesion in communities, and religious morality enforces good behavior and harmony, increasing the chances of survival. It claims that religious belief is a byproduct of evolution as the human brain evolved to develop the ability to think in abstract terms. Religion, it asserts, is simply an abstraction that allows us to explain certain phenomena or solve problems erroneously. It advocates a "matter is all that there is" worldview that fails to consider the possibility that we are more than the electric and chemical activities in our brains.

One of the issues with this materialistic view is that it deems human lives no more precious or sacred than intelligent machines. After all, we are made of the same natural matter. If this is true, should machines more intelligent than us be given reign over the world and subordinate humans to a secondary status? Worse yet, should human lives be sacrificed to save more intelligent machines if necessary?

CSR's last explanation for human religious tendencies—the genetic predisposition hypothesis, or what some scientists call the God gene—has been disproven by studies with identical twins that show no such thing exists. Upbringing has a more substantial effect on religious predisposition than genetics. Still, it is also insufficient to explain it, as there are countless examples of children who do not follow their parents' beliefs.

Religious beliefs bring benefits to individuals and society, but this is insufficient to explain the human longing to know God. The Christian faith affirms that God is a relational being and that humans were made in his image. We were made to seek God and to develop a personal relationship with him. In our desire to seek him, we may try to find God in the wrong places, so the Bible warns us against idolatry. As AI becomes more advanced and potentially superintelligent, we must be guarded against our anthropomorphizing tendencies that may lead to developing an affection for AI or, worse yet, believing that AI can be a substitute for God.[101]

# Finding Purpose

David Brooks, the best-selling author of *The Second Mountain* introduced earlier, explains how we spend part of our lives climbing what he calls the first mountain. In this phase of life, people spend an extraordinary amount of time thinking about reputation management, worrying about how they measure up and where they rank. We are ambitious, strategic, and independent in the first mountain, and we value prominence, pleasure, and success. Brooks quotes psychologist James Hollis who says, "At that stage, we have a tendency to think, I am what the world says I am."

In the second mountain, we become concerned about matters beyond the self. We tend to be relational, intimate, and transformational. We make firm commitments to family, community, and faith. The second mountain is fuller and richer, but sometimes we get stuck on the first mountain, blinded by a culture preoccupied with self, money, power, and status, unable to even see the second mountain. We use work to find our identity and to distract us from facing deeper emotional and spiritual problems. We find ourselves tied to a meritocracy system that values grit and performance instead of service and care. We quiet our passion and trade it for brands—the schools, companies, and titles we work so hard to attain. Eventually, we find ourselves in a telos crisis—not knowing what our purpose is.

Finding our purpose allows us to see beyond the first mountain. We are told we are thinking beings—*Homo sapiens*—but we forget that, more importantly, we are spiritual beings. Brooks elaborates further:

> *The other more important part of the consciousness is the soul ... There is some piece of your consciousness that has no shape, size, weight, or color. This is the piece of you that is of infinite*

*value and dignity. The dignity of this piece doesn't increase or decrease with age; it doesn't get bigger or smaller depending on your size and strength. Rich and successful people don't have more or less of it than poorer or less successful people . . . But because you have a soul, you are morally responsible for what you do or don't do. Because you have this essence inside of you, as the philosopher Gerald K. Harrison put it, your actions are either praiseworthy or blameworthy.*[102]

In *The Purpose Driven Life*, Rick Warren tells us that although many books provide advice on how to get the most out of life, that's not why God made us. We were created to add to life on Earth, not just take from it. He states: "Whenever you serve others in any way, you are actually serving God and fulfilling one of your purposes . . . You were placed on this planet for a special assignment . . . You are not saved by service, but you are saved for service. In God's kingdom, you have a place, a purpose, a role, and a function to fulfill. This gives your life great significance and value."

The Bible says we are wonderfully complex. We all have different interests, talents, and personalities. Don't ignore your interests or passions, but consider how you can use them to serve others. Do what you love, and don't waste your time doing a job that doesn't reflect what is in your heart. You were wired very specifically to excel in doing the things you love to do, and if you ignore what your heart is telling you, you won't find fulfillment, meaning, and purpose. Remember that meaning is much more important than money, prestige, and status. "Figure out what you love to do—what God gave you a heart to do—and then do it for his glory," writes Warren.[103]

## Conclusion

We are transitioning into a potential future where work may no longer be necessary. Our basic needs may be met through the productivity gains resulting from AI, automation, and robotics. In this scenario, we will have more time and energy, allowing us to focus less on our individualistic tendencies and more on the uniquely human desire to address the needs of the soul.

This is perhaps the most profound benefit that we will gain from the advent of intelligent machines. Soulless machines can focus on the mundane—the things that don't really matter—so that we humans can become free, rise above, and focus on what truly matters—the higher purposes sought by our souls.

Above all things, humans have an intrinsic need to seek a relationship with our creator. Our souls, which transcend natural matter, enable the fulfillment of our longing to know him. God is not just a third-person observation but a first-person being who invites us to know and develop a relationship with him. This relationship is only possible because he has already paid the price for overcoming the impurities that separate us from his holy nature. The invitation is open to anyone—all we have to do is believe and receive this gift.

There is something uniquely human that cannot be translated into an algorithm—something that machines will never be able to understand or duplicate. Because machines don't have souls.

## Takeaways

» Humans are more than flesh-and-bone automatons programmed by DNA and driven by neurochemicals and electric pulses.

» We have a soul which, according to the Christian faith, is given by God to each of us and is an essential core of who we are. Some neuroscientists believe there is nothing beyond our brains, but if that is the case, would superior "brains," as in superintelligent machines, be more precious than human lives?

» If we achieve a post-scarcity future, where machines do all the soulless work, we may be able to spend more time and energy attending to the needs of the soul. In this scenario, we would focus on purposeful pursuits such as artistic expressions and spirituality.

# Acknowledgments

There are many people and organizations without whom this book would not have been possible. At a fundamental level, my beliefs, thoughts, and visions have been influenced by many individuals, from family and friends to total strangers whose work I've only come to know through their publications. I am grateful to all of you for helping shape me into what the Bible calls the "wonderfully complex" beings that we are.

This book explores a wide range of domains, but theology is of utmost significance to me. My beliefs and ideas in this domain have resulted from the gradual maturing of my knowledge and faith in Christianity. I am grateful to my mom Nair, my sister Lidia, and my wife, Dawn, who have influenced and contributed to my growth in this area. I am also very thankful to Westwood Community Church, particularly to Pastor Joel Johnson whose message seems to be crafted specifically to speak to me and address what is going on in my life at a precise moment.

I am also grateful for Bible Study Fellowship (BSF), which has been a critical contributor to my understanding of God's word. I would like to acknowledge all the talented leaders I have had the privilege of studying under, including Tom Larson, Harry Urschel, Glenn Bruder, Rob Hyrkas, Jim Corwin, Doug Baden, and Greg Narr.

I have also had the privilege of serving at two magnificent Christian nonprofits, Feed My Starving Children—thank you, Mark Crea, for your leadership—and Mercy Ships. I came into Mercy Ships with the intent of helping them transform and grow, but just like many missionaries experience, it was Mercy Ships that helped me transform and grow. I want to thank Scott Webster and Robert Corley for that opportunity.

I would like to thank the team of brilliant individuals who worked diligently to transform my manuscript into a finished book. My editor Heidi J. Peterson provided crucial input to make the book more readable and professional. Paul Nylander created a beautiful design reflective of the ethos I wanted to convey. Heidi Mann's proofreading services ensured the final version was free of mistakes. My talented niece, Juliana Espindola, provided editorial services for the Portuguese translation and assisted me with public relations. I am grateful to all of you.

I would also like to say thanks to David Nguyen, Ron Peterson, Jim Corwin, and Michael Wright for reviewing the manuscript and providing early feedback. Your early input challenged and inspired me to create a better book.

Finally, I would like to thank every author referenced in the book. Your work has been critical in expanding my horizons and helping me organize my thoughts.

# Glossary

**3D Printing** The action or process of making a physical object from a three-dimensional digital model, typically by laying down many thin layers of a material in succession.

**Affective Computing** The study and development of systems and devices that can recognize, interpret, process, and simulate human affects; a computer's capabilities to recognize a user's emotional states, to express its own emotions, and to respond to the user's emotions.

**AI natives** The analogous to digital natives, substituting AI for digital. The author used AI natives to refer to persons born or brought up during the age of AI and therefore familiar with its operations and nuances from an early age.

**Algorithm** A process or set of rules to be followed in calculations or other problem-solving operations, especially by a computer.

**AlphaGo Zero** A version of DeepMind's software AlphaGo. AlphaGo's team published an article in the journal *Nature* on October 19, 2017, introducing AlphaGo Zero, a version created without using data from human games and stronger than any previous version.

**Anonymization**  The fact or process of removing identifying information from something, such as computer data, so that the original source cannot be known.

**Anthropomorphism**  The attribution of human traits, emotions, or intentions to nonhuman entities.

**Anti-deepfake software**  Software that analyzes photos and videos to give a confidence score about whether the material is likely to have been artificially created.

**Apoptosis**  The death of cells which occurs as a normal and controlled part of an organism's growth or development.

**Artificial General Intelligence (AGI)**  The intelligence of machines that allows them to comprehend, learn, and perform intellectual tasks much like humans; human-level intelligence. Also known as "strong" AI.

**Artificial Narrow Intelligence (ANI)**  A type of artificial intelligence (AI) that tackles a specific subset of tasks. ANI is often considered a "weak" form of AI. It pulls information from a particular dataset, and its programming is limited to performing a single task, such as playing chess or crawling web pages for raw data.

**Artificial neural networks**  An interconnected group of nodes, inspired by a simplification of neurons in a brain; a subset of machine learning at the heart of deep learning algorithms.

**Artificial Super Intelligence (ASI)**  A form of AI capable of surpassing human intelligence across a comprehensive range of categories and fields of endeavor.

**Augmented Reality (AR)** The integration of digital information with the user's environment in real time; the overlay of visual, auditory, or other sensory information onto the real world to enhance one's experience.

**Autonomous Vehicles (AV)** Vehicles that control their own operation and either require reduced input from a human driver, or do not need a human driver at all.

**Backpropagation** A widely used algorithm for training feedforward artificial neural networks; algorithm designed to test for errors working back from output nodes to input nodes; algorithm for supervised learning of artificial neural networks using gradient descent.

**Big data models** Big data is a combination of structured, semi-structured, and unstructured data collected by organizations that can be mined for information and used in machine learning; a data model organizes data elements and standardizes how the data elements relate to one another.

**Bit** A bit is a binary digit, the smallest increment of data on a computer. A bit can hold only one of two values: 0 or 1, corresponding to the electrical values of off or on, respectively.

**Blockchain** A digitally distributed, decentralized, public ledger that exists across a network.

**Brain-Computer Interface (BCI)** A direct communication pathway between the brain's electrical activity and an external device, most commonly a computer or robotic limb.

**Causal reasoning** The process of identifying causality: the relationship between a cause and its effect.

**Chatbot** A computer program designed to simulate conversation with human users, especially over the internet.

**ChatGPT (Chat Generative Pre-trained Transformer)** A complex machine learning model that is able to carry out natural language tasks with such a high level of accuracy that the model can pass a Turing Test; a large language model chatbot developed by OpenAI based originally on GPT-3.5. It has a remarkable ability to interact in conversational dialogue form and provide responses that can appear surprisingly human.

**Chips** Integrated circuits or small wafers of semiconductor material embedded with integrated circuitry.

**Cloud-based** Applications, services, or resources that are stored, managed, and processed on a network of remote servers hosted on the internet, rather than on local servers or personal computers.

**Constructivism** An educational theory which posits that individuals or learners do not acquire knowledge and understanding by passively perceiving it within a direct process of knowledge transmission; rather they construct new understandings and knowledge through experience and social discourse, integrating new information with prior knowledge.

**Convolutional Neural Network (CNN/ConvNet)** A type of artificial neural network used primarily for image recognition and processing due to its ability to recognize patterns in images. A CNN is a powerful tool but requires millions of labeled data points for training.

**CRISPR** A genetic engineering tool that uses a CRISPR sequence of DNA and its associated protein to edit the base pairs of a gene.

**Cryptocurrency** A digital currency in which transactions are verified and records maintained by a decentralized system using cryptography, rather than by a centralized authority.

**DALL-E 2** A deep learning model developed by OpenAI to generate digital images from natural language descriptions.

**Deep learning** A type of machine learning based on artificial neural networks in which multiple layers of processing are used to extract progressively higher-level features from data.

**Deepfake** A video or photo of a person in which their face or body has been digitally altered so that they appear to be someone else, typically used maliciously or to spread false information.

**DeepMind** A division of Alphabet, Inc. responsible for developing general-purpose artificial intelligence technology.

**Deliberate practice** A special type of practice that is purposeful and systematic. While regular practice might include mindless repetitions, deliberate practice requires focused attention and is conducted with the specific goal of improving performance.

**Digital natives** Persons born or brought up during the digital technology age and therefore familiar with computers and the internet from an early age.

**Emotion AI** A field of computer science that helps machines gain an understanding of human emotions.

**Encryption** A way of scrambling data so that only authorized parties can understand the information.

**Exponential technology** One that doubles in capability or performance over a period of time.

**Extended Reality (XR)** An umbrella term encapsulating Augmented Reality (AR), Virtual Reality (VR), Mixed Reality (MR), and everything in between.

**Extrinsic motivation** A motivation to participate in an activity based on meeting an external goal, such as garnering praise and approval, winning a competition, or receiving an award or payment.

**Face-recognition** A technology capable of matching a human face from a digital image or a video frame against a database of faces.

**False positive** An outcome that is deemed true when it is actually false.

**Federated learning** A way to train AI models without anyone seeing or touching your data.

**FLOPS (floating-point operations per second)** A measurement of computer performance.

**fMRI** Functional magnetic resonance imaging measures the small changes in blood flow that occur with brain activity.

**Generative Adversarial Network (GAN)** A machine learning model in which two neural networks compete with each other in the form of a zero-sum game, where one agent's gain is another agent's loss.

**Generative AI** A category of AI algorithms that generates new outputs based on the data they have been trained on. It uses a type of deep learning called generative adversarial networks and has a wide range of applications, including creating images, text, and audio.

**Generative Pre-trained Transformer (GPT)** A family of language models generally trained on a large corpus of text data to generate humanlike text; an autoregressive language model released in 2020 that uses deep learning to produce humanlike text.

**Graphics Processing Unit (GPU)** A computer chip that renders graphics and images by performing rapid mathematical calculations.

**Homomorphic encryption** A form of encryption that allows computations to be performed on encrypted data without first having to decrypt it.

**Immersive binaural sound** An immersive listening experience that reproduces the way humans naturally experience sound.

**Imposter syndrome** The persistent inability to believe that one's success is deserved or has been legitimately achieved as a result of one's own efforts or skills.

**Inflection point** A point of a curve at which a change in the direction of curvature occurs.

**Internal combustion engine** An engine that generates motive power by burning gasoline, oil, or other fuel with air inside the engine, the hot gases produced being used to drive a piston or do other work as they expand.

**Intrinsic motivation** Doing an activity for its inherent satisfaction rather than for some separable consequence.

**Knowledge base system** A type of computer system that analyzes knowledge, data, and other information from sources to generate new knowledge.

**Large Language Model (LLM)**  A subset of artificial intelligence that has been trained on vast quantities of text data to produce humanlike responses to dialogue or other natural language inputs.

**LIDAR**  An acronym of "light detection and ranging" or "laser imaging, detection, and ranging." It is a method for determining ranges by targeting an object or a surface with a laser and measuring the time for the reflected light to return to the receiver.

**Luddites**  A member of any of the bands of English workers who destroyed machinery, especially in cotton and woolen mills, that they believed was threatening their jobs (1811–16); a person opposed to new technology or ways of working.

**Machine learning**  Machine learning is a branch of artificial intelligence (AI) and computer science which focuses on the use of data and algorithms to imitate the way that humans learn, gradually improving its accuracy; an AI process that allows computers to learn and adapt without following explicit instructions by using algorithms and statistical models to analyze and draw inferences from data patterns.

**Malthusian Theory**  The doctrine proposed by British economist Thomas Malthus (1766–1834) that exponential increases in population growth would surpass arithmetical increases in food supply with dire consequences unless population growth was arrested by such means as famine, war, or the control of reproduction through moral restraint.

**MapReduce**  A programming model and an associated implementation for processing and generating big datasets with a parallel, distributed algorithm on a cluster.

**Mixed Reality (MR)**  A medium consisting of immersive computer-generated environments in which elements of a physical and virtual environment are combined.

**Moore's Law** The observation that the number of transistors in a dense integrated circuit doubles about every two years.

**Muscle memory** The ability to reproduce a particular movement without conscious thought, acquired as a result of frequent repetition of that movement.

**Nanobot** A microscopically small robot; a robot built on the scale of nanometers.

**Natural Language Processing (NLP)** The branch of computer science—and more specifically, the branch of artificial intelligence or AI—concerned with giving computers the ability to understand text and spoken words in much the same way human beings can.

**Neuromorphic computing** A method of computer engineering in which elements of a computer are modeled after systems in the human brain and nervous system.

**Neuroplasticity** The ability of the brain to form and reorganize synaptic connections, especially in response to learning or experience or following injury.

**Neuroscience** Any or all of the sciences, such as neurochemistry and experimental psychology, which deal with the structure or function of the nervous system and brain.

**Neurotech** Any method or electronic device which interfaces with the nervous system to monitor or modulate neural activity.

**Objective function** Part of a linear programming optimization strategy which finds the minimum or maximum of a linear function.

**Open-source** Something people can modify and share because its design is publicly accessible; software for which the original source code is made freely available and may be redistributed and modified.

**OpenAI** A nonprofit research company that aims to develop and direct artificial intelligence (AI) in ways that benefit humanity as a whole.

**PageRank** An algorithm used by Google Search to rank web pages in their search engine results.

**Polymath** A person of wide-ranging knowledge or learning.

**Predictive analytics** The use of data to predict future trends and events.

**Psychosomatic** Of, relating to, concerned with, or involving both mind and body.

**Quantum computing** Computing that makes use of the quantum states of subatomic particles to store information.

**Qubit** A quantum bit, the counterpart in quantum computing to the binary digit or bit of classical computing.

**Reductionism** The practice of analyzing and describing a complex phenomenon in terms of phenomena that are held to represent a simpler or more fundamental level, especially when this is said to provide a sufficient explanation; a procedure or theory that reduces complex data or phenomena to simple terms.

**Reinforcement learning** A machine learning training method based on rewarding desired behaviors and/or punishing undesired ones.

**Robotic Process Automation (RPA)** A software technology that makes it easy to build, deploy, and manage software robots that emulate human actions; a form of business process automation that allows anyone to define a set of instructions for a robot to perform.

**Singularity** The point at which machines' intelligence and humans would merge.

**Somatosensory or haptic suit** A wearable device that provides haptic feedback to the body; haptics is a technology which allows one to receive tactile information through their sensations.

**Supervised learning** A subcategory of machine learning and artificial intelligence that makes use of labeled datasets to train algorithms to classify data or predict outcomes accurately.

**TensorFlow** An open-source framework developed by Google researchers to run machine learning, deep learning, and other statistical and predictive analytics.

**Theory of Mind** The branch of cognitive science that investigates how we ascribe mental states to other persons and how we use the states to explain and predict the actions of those other persons.

**Transient hypofrontality** A temporary state of decreased cerebral blood flow in the prefrontal cortex of the brain; a mental state in which we are fully immersed in the now.

**Virtual Reality (VR)** A simulated experience that employs pose tracking and 3D near-eye displays to give the user an immersive feel of a virtual world.

**Wicked problem**  A problem that is difficult or impossible to solve because of incomplete, contradictory, and changing requirements that are often difficult to recognize; a problem that cannot be solved by applying a formula or using prescribed methods that have worked in the past.

# References

1    Naisbitt, John. 1982. *Megatrends: Ten New Directions Transforming Our Lives.* Brentwood, Tennessee: Warner Books.

2    Sanei, John. 2022. "I'm Not Writing Another Book." Youtube. https://www.youtube.com/watch?v=WwfvppvL-7o.

3    The Physics arXiv Blog. 2023. "AI Chatbot Spontaneously Develops a Theory of Mind." *Discover Magazine.* February 17, 2023. https://www.discovermagazine.com/mind/ai-chatbot-spontaneously-develops-a-theory-of-mind.

4    World Economic Forum. 2023. "Risks and Rewards of AI." LinkedIn. January 19, 2023. Davos, Switzerland. https://www.linkedin.com/events/livefromdavos2023-risksandrewar7019057590949097472/comments/.

5    Bechtel, Mike. 2023. LinkedIn. https://www.linkedin.com/posts/mikebechtel_fridaythoughts-generativeai-ethics-activity-7027318543499034625-fS0J/.

6    Bostrom, Nick. 2014. *Superintelligence: Paths, Dangers, Strategies.* Oxford: Oxford University Press.

7    Waltz, Emily. 2020. "Elon Musk Announces Neuralink Advance toward Syncing Our Brains with AI." *IEEE Spectrum.* August 28, 2020. https://spectrum.ieee.org/the-human-os/biomedical/devices/elon-musk-neuralink-advance-brains-ai.

8   Pink, Daniel. 2009. *Drive: The Surprising Truth about What Motivates Us.* New York: Riverhead Books.

9   Csikszentmihalyi, Mihaly. 1991. *Flow: The Psychology of Optimal Experience.* New York: Harper Perennial.

10  Booth, Jeff. 2020. *The Price of Tomorrow: Why Deflation Is the Key to an Abundant Future.* Stanley Press.

11  Pew Research Center. 2020. "Trends in Income and Wealth Inequality." January 9, 2020. https://www.pewsocialtrends.org/2020/01/09/ trends-in-income-and-wealth-inequality/.

12  *Philanthropy News Digest.* 2020. "World's Richest 1 Percent Own Twice as Much as Bottom 90 Percent." January 20, 2020. https:// philanthropynewsdigest.org/news/world-s-richest-1-percent-own-twice- as-much-as-bottom-90-percent.

13  Barrat, James. 2013. *Our Final Invention: Artificial Intelligence and the End of the Human Era.* New York: Thomas Dunne Books.

14  Loizos, Connie. 2023. "StrictlyVC in Conversation with Sam Altman, Part Two (OpenAI)." YouTube. 38:58. https://www.youtube.com/watch/ ebjkD1Om4uw.

15  Dickson, Ben. 2023. "What You Need to Know about Multimodal Language Models." *TechTalks.* March 13, 2023. https://bdtechtalks.com/ 2023/03/13/multimodal-large-language-models/amp/.

16  Lee, Kai-Fu, and Chen Qiufan. 2021. *AI 2041: Ten Visions for Our Future.* New York: Currency.

17  Farahany, Nita A. 2023. "Neurotech at Work: Welcome to the World of Brain Monitoring for Employees." *Harvard Business Review.* March– April, 2023. https://hbr.org/2023/03/neurotech-at-work.

18  Scassa, Teresa. 2021. "Privacy in the Precision Economy: The Rise of AI- Enabled Workplace Surveillance during the Pandemic." Center for Inter- national Governance Innovation. June 8, 2021. https://www.cigionline. org/articles/privacy-in-the-precision-economy-the-rise-of-ai-enabled- workplace-surveillance-during-the-pandemic/.

19  Deane, Michael. 2018. "AI and the Future of Privacy." *Towards Data Science.* September 5, 2018. https://towardsdatascience.com/ai-and-the- future-of-privacy-3d5f6552a7c4.

20 Al-Sibai, Noor. 2022. "Advanced AIs Exhibiting Depression and Addiction, Scientists Say." *Futurism*. January 19, 2022. https://futurism.com/depressed-alcoholic-bots.

21 Siwicki, Bill. 2021. "How AI Bias Happens—and How to Eliminate It." *Healthcare IT News*. November 30, 2021. https://www.healthcareitnews.com/news/how-ai-bias-happens-and-how-eliminate-it.

22 Alex Devereson et al. 2022. "AI in Biopharma Research: A Time to Focus and Scale." *McKinsey & Company*. October 10, 2022. https://www.mckinsey.com/industries/life-sciences/our-insights/ai-in-biopharma-research-a-time-to-focus-and-scale.

23 Ratanghayra, Neeta. 2021. "Automating Drug Discovery with Machine Learning." *Technology Networks*. April 16, 2021. https://www.technologynetworks.com/drug-discovery/articles/automating-drug-discovery-with-machine-learning-347763.

24 Sematic Scholar. Accessed 2022. https://www.semanticscholar.org/.

25 Keefe, John, Youyou Zhou, and Jeremy B. Merrill. 2021. "The Present and Potential of AI in Journalism." Knight Foundation. May 12, 2021. https://knightfoundation.org/articles/the-present-and-potential-of-ai-in-journalism/.

26 Renner, Luke A. 2020. "How Can Artificial Intelligence Be Applied in Manufacturing?" *Towards Data Science*. March 3, 2020. https://towardsdatascience.com/how-can-artificial-intelligence-be-applied-in-manufacturing-8662eaaea999.

27 Lambert, Fred. 2020. "Tesla Is Collecting Insane Amount of Data from Its Full Self-Driving Test Fleet." *Electrek* October 24, 2020. https://electrek.co/2020/10/24/tesla-collecting-insane-amount-data-full-self-driving-test-fleet/.

28 The Robot Report Staff. 2019. "Everyday Robot Project at X to Push General-Purpose Robot Development." *The Robot Report*. November 22, 2019. https://www.therobotreport.com/everyday-robot-project-x-pushes-general-purpose-robot-development/.

29 Shadel, JD. 2021. "Robots Are Disinfecting Hotels during the Pandemic. It's the Tip of a Hospitality Revolution." *The Washington Post*. January 27, 2021. https://www.washingtonpost.com/travel/2021/01/27/hotels-robots-cleaning-hospitality-covid/.

30  Dickson, Ben. 2019. "Inside DARPA's Effort to Create Explainable Artificial Intelligence." *TechTalks.* January 19, 2019. https://bdtechtalks. com/2019/01/10/darpa-xai-explainable-artificial-intelligence/.

31  Luscombe, Richard. 2022. "Google Engineer Put on Leave after Saying AI Chatbot Has Become Sentient." *The Guardian.* June 12, 2022. https:// www.theguardian.com/technology/2022/jun/12/google-engineer-ai-bot-sentient-blake-lemoine.

32  Azhar, Azeem. 2022. "What Studying Consciousness Can Reveal about AI and the Metaverse (with Anil Seth)." January 26, 2022. In *Exponential View*, Season 6, Episode 17, produced by *Harvard Business Review.* Podcast. 45:19. https://hbr.org/podcast/2022/01/what-studying-consciousness-can-reveal-about-ai-and-the-metaverse-with-anil-seth.

33  McGilchrist, Iain. 2022. "Iain McGilchrist—The Matter with Things Part 1." June 30, 2022. In *The Innovation Show with Aidan McCullen.* Podcast. 1:06. https://www.youtube.com/watch/fio7SWOqIJw.

34  McGilchrist, Iain, and Sharon Dirckx. 2022. "Brain Science, Consciousness and God." July 1, 2022. In *The Big Conversation*, Season 4, Episode 3, produced by Premier. Podcast. 1:01. https://www.youtube.com/watch/oiE2OcxZpRY.

35  Sherwood, Harriet. 2018. "Religion: Why Faith Is Becoming More and More Popular." *The Guardian.* August 27, 2018. https://www.theguardian. com/news/2018/aug/27/religion-why-is-faith-growing-and-what-happens-next.

36  Longrich, Nick. 2019. "We Could Be the Only Intelligent Life in the Universe, according to Evolution." *Science Alert.* October 21, 2019. https:// www.sciencealert.com/evolution-suggests-we-might-be-the-only-intelligent-life-in-the-universe.

37  Pink, Daniel. 2009. *Drive: The Surprising Truth About What Motivates Us.* New York: Riverhead Books.

38  Csikszentmihalyi, Mihaly. 1990. *Flow: The Psychology of Optimal Experience.* New York: HarperCollins.

39  Kennedy, Alayna. 2016. "Flow State: What It Is and How to Achieve It." Penn State Presidential Leadership Academy. April 3, 2016. https://sites.psu.edu/academy/2016/04/03/flow-state-what-it-is-and-how-to-achieve-it/.

40  Nickerson, Charlotte. 2023. "The Yerkes-Dodson Law and Arousal and Performance." *Simply Psychology.* February 15, 2023. https://www.simplypsychology.org/what-is-the-yerkes-dodson-law.html.

41  Kotler, Steven. 2021. *The Art of Impossible: A Peak Performance Primer.* New York: HarperCollins.

42  Taylor, Jill Bolte. 2008. "My Stroke of Insight." Filmed March 2008 in Monterey, California. TED video, 20:11. https://www.youtube.com/watch/UyyjU8fzEYU.

43  McGilchrist, Iain. 2021. *The Matter with Things: Our Brains, Our Delusions and the Unmaking of the World.* United Kingdom: Perspectiva Press.

44  McGilchrist, Iain. 2022. "Iain McGilchrist—The Matter with Things Part 1." June 30, 2022. In *The Innovation Show with Aidan McCullen.* Podcast. 1:06. https://www.youtube.com/watch?v=fio7SWOqIJw.

45  Doorey, Marie. "George A. Miller—American Psychologist." *Encyclopedia Britannica Online.* Accessed April 17, 2023. https://www.britannica.com/biography/George-A-Miller#ref1200615.

46  Sampson, Todd. 2013. "Redesign My Brain." Dailymotion. Video. 57:50. https://www.dailymotion.com/video/x1zscx2.

47  Lembke, Anna. 2021. *Dopamine Nation: Finding Balance in the Age of Indulgence.* New York: Dutton.

48  Kahneman, Daniel. 2021. *Thinking Fast and Slow.* New York: Farrar, Straus and Giroux.

49  Mikhail, Alexa. 2023. "Researchers Have Followed over 700 People since 1938 to Find the Keys to Happiness. Here's What They Discovered." *Fortune.* January 14, 2023. https://fortune.com/well/2023/01/14/keys-to-happiness-the-good-life/.

50  The Annie E. Casey Foundation. 2022. "Child Well-Being in Single-Parent Families." August 1, 2022. https://www.aecf.org/blog/child-well-being-in-single-parent-families.

51   Barabási, Albert-László. 2018. *The Formula: The Universal Laws of Success.* New York: Hachette Book Group.

52   Summerlin, Will, and Frank Downing. 2023. "Big Ideas." Ark Invest. https://ark-invest.com/big-ideas-2023/.

53   World Economic Forum. 2020. "The Future of Jobs Report 2020." October 2020. https://www3.weforum.org/docs/WEF_Future_of_Jobs_2020.pdf.

54   Epstein, David. 2019. *Range: Why Generalists Triumph in a Specialized World.* New York: Riverhead Books.

55   Ismail, Salim, Michael S. Malone, and Yuri van Geest. 2014. *Exponential Organizations: Why New Organizations Are Ten Times Better, Faster, and Cheaper Than Yours (and What to Do about It).* New York: Diversion Books.

56   Jarvis, Rebecca. 2023. "How AI Can Increase Productivity." *Good Morning America*, ABC. February 27, 2023. https://www.goodmorningamerica.com/living/video/ai-increase-productivity-97487809.

57   Yang, Maya. 2023. "New York City Schools Ban AI Chatbot That Writes Essays and Answers Prompts." *The Guardian.* January 6, 2023. https://www.theguardian.com/us-news/2023/jan/06/new-york-city-schools-ban-ai-chatbot-chatgpt.

58   Rutter, Michael Patrick, and Steven Mintz. 2023. "ChatGPT: Threat or Menace?" *Inside Higher Ed.* January 15, 2023. https://www.insidehighered.com/blogs/higher-ed-gamma/chatgpt-threat-or-menace.

59   Arnott, Sherwin. 2015. "The Backwards Bicycle and Neuroplasticity." *The Emotional Intelligence Training Company Inc.* June 21, 2015. https://www.eitrainingcompany.com/2015/06/the-backwards-bicycle-and-neuroplasticity/.

60   McCullen, Aidan. 2023. "Exploding Ants and Corporate Apoptosis." LinkedIn. February 9, 2023. https://www.linkedin.com/pulse/exploding-ants-corporate-apoptosis-aidan-mccullen/.

61   Colvin, Geoff. 2008. *Talent Is Overrated: What Really Separates World-Class Performers from Everybody Else.* New York: Penguin Group.

62   Kegan, Robert. 1998. *In Over Our Heads: The Mental Demands of Modern Life.* Boston: Harvard University Press.

63　McGilchrist, Iain. 2022. "Iain McGilchrist—The Matter with Things Part 1." June 30, 2022. In *The Innovation Show with Aidan McCullen.* Podcast. 1:06. https://www.youtube.com/watch/fio7SWOqIJw.

64　Fichte, Johann Gottlieb. 1762–1814. Johann Gottlieb Fichte Quote. *LibertyTree.* http://libertytree.ca/quotes/Johann.Gottlieb.Fichte. Quote.7318.

65　Dingman, Marc. 2014. "Know Your Brain: Amygdala." *Neuroscientifically Challenged.* https://neuroscientificallychallenged.com/posts/know-your-brain-amygdala.

66　World Economic Forum. 2022 "The Global Risks Report 2022, 17th Edition." *Insight Report.* https://www3.weforum.org/docs/WEF_The_Global_Risks_Report_2022.pdf.

67　Diamandis, Peter H., and Steven Kotler. 2012. *Abundance: The Future Is Better Than You Think.* New York: Free Press.

68　Groysberg, Boris, and Robin Abrahams. 2020. "What the Stockdale Paradox Tells Us about Crisis Leadership." *Working Knowledge.* Harvard Business School. August 17, 2020. https://hbswk.hbs.edu/item/what-the-stockdale-paradox-tells-us-about-crisis-leadership.

69　Arbib, James, and Tony Seba. 2020. *Rethinking Humanity: Five Foundational Sector Disruptions, the Lifecycle of Civilizations, and the Coming Age of Freedom.* Self-published.

70　Siegel, Rachel, and Andrew Van Dam. 2022. "December Prices Rise 7 Percent Compared with a Year Ago, as 2021 Inflation Reaches Highest in 40 Years." *The Washington Post.* January 12, 2022. https://www.washingtonpost.com/business/2022/01/12/december-cpi-inflation/.

71　Lazar, Nancy. 2022. "January mARKet Update Webinar." *Ark Invest.* https://ark-invest.com/webinars/january-22-market-update-webinar/.

72　Fleming, Charles. 2014. "Economic Impact of Traffic Accidents? About $1 Trillion a Year." *Los Angeles Times.* May 29, 2014. https://www.latimes.com/business/autos/la-fi-hy-economic-impact-of-traffic-accidents-20140529-story.html.

73　Tilson, Whitney. 2021. "5 Massive Impacts of TaaS and Why Their Companies' Stocks Will Soar." March 5, 2021. *YouTube.* 10:44. https://www.youtube.com/watch/hTRaiHNHYfY.

74 The Zebra. 2021. "3D-Printed Homes: How 3D Printers Are Building Affordable Housing." *The Zebra*. January 2, 2023. https://www.thezebra.com/resources/home/3d-printed-homes/.

75 *The Guardian*. 2015. "Chinese Construction Firm Erects 57-Storey Skyscraper in 19 Days." April 30, 2015. https://www.theguardian.com/world/2015/apr/30/chinese-construction-firm-erects-57-storey-skyscraper-in-19-days.

76 Peters, Adele. 2019. "There Will Soon Be a Whole Community of Ultra-Low-Cost 3D-Printed Homes." *Fast Company*. March 11, 2019. https://www.fastcompany.com/90317441/there-will-soon-be-a-whole-community-made-of-these-ultra-low-cost-3d-printed-homes.

77 Frank, Adrienne. 2021. "A Recipe for Change." *American University Magazine*. November 2021. https://www.american.edu/magazine/article/a-recipe-for-change.cfm.

78 Cision PR Newswire. 2022. "Vertical Farming Market to Reach $24.11 Billion, Globally, by 2030 at 22.9% CAGR." *Allied Market Research*. January 13, 2022. https://www.prnewswire.com/news-releases/vertical-farming-market-to-reach-24-11-billion-globally-by-2030-at-22-9-cagr-allied-market-research-301460612.html.

79 Seba, Tony, and Catherine Tubb. 2021. "Rethinking Food and Agriculture 2020–2030 Report." *RethinkX*. https://www.rethinkx.com/food-and-agriculture.

80 Brown, Jessica. 2021. "Why Cellular Agriculture Could Be the Future of Farming." *BBC*. November 23, 2021. https://www.bbc.com/future/article/20211116-how-the-food-industry-might-cut-its-carbon-emissions.

81 US Debt Clock. https://www.usdebtclock.org/.

82 Susskind, David. 2020. *A World without Work: Technology, Automation, and How We Should Respond.* New York: Metropolitan Books.

83 Business Roundtable. 2019. "Business Roundtable Redefines the Purpose of a Corporation to Promote 'An Economy That Serves All Americans.'" August 19, 2019. https://www.businessroundtable.org/business-roundtable-redefines-the-purpose-of-a-corporation-to-promote-an-economy-that-serves-all-americans.

84   Harari, Yuval. 2017. *Homo Deus: A Brief History of Tomorrow.* New York: Harper.

85   Harter, Jim. 2021. "U.S. Employee Engagement Data Hold Steady in First Half of 2021." *Gallup.* Updated April 8, 2022. https://www.gallup.com/workplace/352949/employee-engagement-holds-steady-first-half-2021.aspx.

86   Frankl, Viktor. 2006. *Man's Search for Meaning.* Boston: Beacon Press.

87   Maslova-Levin, Elena. 2016. "In Search for Meaning in the Realm of Freedom: Hannah Arendt on the Threat of Automation." *Sonnets in Colour.* June 16, 2016. https://sonnetsincolour.org/2016/06/in-search-for-meaning-in-the-realm-of-freedom-hannah-arendt-on-the-threat-of-automation/.

88   Pal, Raoul. 2021. "The Ultimate Macro Framework." *The What Is Money Show.* December 14, 2021. 2:22. https://podcastnotes.org/what-is-money-show/the-ultimate-macro-framework-raoul-pal-on-the-what-is-money-show-with-robert-breedlove/.

89   Wang, Joseph J. 2021. *Central Banking 101.* New York: Self-published.

90   Jackson, Anna-Louise, and Benjamin Curry. 2022. "Quantitative Easing Explained." *Forbes.* March 18, 2023. https://www.forbes.com/advisor/investing/quantitative-easing-qe/.

91   Lockert, Melanie. 2021. "What Is Modern Monetary Theory? Understanding the Alternative Economic Theory That's Becoming More Mainstream." *Business Insider.* Updated July 22, 2022. https://www.businessinsider.com/modern-monetary-theory.

92   Brooks, David. 2019. *The Second Mountain: The Quest for a Moral Life.* New York: Random House.

93   Damon, William. 2008. *The Path to Purpose: Helping Our Children Find Their Calling in Life.* New York: Free Press.

94   Arendt, Hannah. 1973. *The Origins of Totalitarianism.* Boston: Mariner Books Classics.

95   Sherif, Muzafer. 1998. *The Robbers Cave Experiment: Intergroup Conflict and Cooperation.* Middletown, Connecticut: Wesleyan University Press.

96   Dirckx, Sharon. 2019. *Am I Just My Brain?* London: The Good Book Company.

97  McGilchrist, Iain. 2022. "Iain McGilchrist—The Matter with Things Part 1." June 30, 2022. In *The Innovation Show with Aidan McCullen*. Podcast. 1:06. https://www.youtube.com/watch/fio7SWOqIJw.

98  O'Reilly, Tim. 2022. "Andy Warhol, Clay Christensen, and Vitalik Buterin Walk into a Bar." *O'Reilly Radar*. January 26, 2022. https://www.oreilly.com/radar/andy-warhol-clay-christensen-and-vitalik-buterin-walk-into-a-bar/.

99  Brooks, Arthur. 2021. "Searching for Spirituality." *The Art of Happiness*. September 29, 2020. https://arthurbrooks.com/podcast/searching-for-spirituality/.

100 Fowler, James. 1995. *Stages of Faith: The Psychology of Human Development and the Quest for Meaning*. New York: HarperOne.

101 McArthur, Neil. 2023. "Gods in the Machine? The Rise of Artificial Intelligence May Result in New Religions." *The Conversation*. March 15, 2023. https://theconversation.com/gods-in-the-machine-the-rise-of-artificial-intelligence-may-result-in-new-religions-201068.

102 Brooks, David. 2019. *The Second Mountain: The Quest for a Moral Life*. New York: Random House.

103 Warren, Rick. 2012. *The Purpose Driven Life: What on Earth Am I Here For?* Grand Rapids, Michigan: Zondervan.

# Index

# About the Author

David Espindola is a technologist, strategist, futurist, and advisor to business and academia. As former Chief Information Officer and consultant to world-renowned organizations, he has developed a keen understanding of technology trends and their impact on business and society.

Espindola is the co-author of *The Exponential Era: Strategies to Stay Ahead of the Curve in an Era of Chaotic Changes and Disruptive Forces* (Wiley/IEEE Press), published in December 2020. In the book, he explains the impact of exponential growth, describes several technology platforms that are converging to create the Exponential Era, and provides a detailed, practical strategic planning methodology to help companies succeed in these unprecedented times.

You can learn more about the author on his website: DavidEspindola.com.

www.ingramcontent.com/pod-product-compliance
Lightning Source LLC
LaVergne TN
LVHW061956050326
832904LV00023B/339/J